Aran Paulus
Antje Klockow-Beck

Analysis of Carbohydrates
by Capillary Electrophoresis

CHROMATOGRAPHIA CE Series

Edited by Kevin D. Altria, Glaxo Wellcome R&D, UK

There are currently a number of general textbooks covering Capillary Electrophoresis where information on a range of applications and techniques can be found. Readers who are interested in a specific area of CE struggle to find truly comprehensive treatments of their areas of interest. The CHROMATOGRAPHIA CE series has been established to allow comprehensive books to be produced covering individual topics. The books are written by well known authors in their specialist application areas and cover CE topics such as DNA analysis, analysis of pharmaceuticals, chiral separations, MECC, carbohydrate analysis, biomedical applications and troubleshooting in CE.

- **Volume 1**: C. Heller (Ed.), Analysis of Nucleic Acids by Capillary Electrophoresis

- **Volume 2**: K. D. Altria, Analysis of Pharmaceuticals by Capillary Electrophoresis

- **Volume 3**: A. Paulus / A. Klockow-Beck, Analysis of Carbohydrates by Capillary Electrophoresis

Aran Paulus
Antje Klockow-Beck

Analysis of Carbohydrates by Capillary Electrophoresis

vieweg

http://www.vieweg.de

Produced by Lengericher Handelsdruckerei, Lengerich

ISBN-13: 978-3-322-85022-5   e-ISBN-13: 978-3-322-85020-1
DOI: 10.1007/978-3-322-85020-1

Preface

Carbohydrates are ubiquitous. They come as monomers, oligomers and polymers with a startling variety of chemical, physical and material properties. They affect almost every aspect of human live. For example they are an essential part of our nutrition. Cellulose, a carbohydrate polymer, acts as a central material in civilization: as building material for housing and as paper for written communication. Scientifically, the interest in carbohydrates ranges from structural elucidation of simple carbohydrates to fundamental biochemical processes such as photosynthesis and glycolysis. Always, a deeper understanding of a particular aspect of carbohydrate chemistry was accompanied by new methods of measuring and characterizing this class of molecules.

During the last decades a better understanding of the roles of carbohydrates in biochemical pathways developed. It turned out, that carbohydrates attached to proteins in glycoproteins are involved in a number of molecular recognition and targeting processes. This lead to a new research field: *Glycobiology*. However, unlike for proteins or nucleic acids, there is at this time no consistent theory how chemical composition and structure controls function and inherent information. In order to develop the database, on which new theories and scientific hypothesis are founded, analytical methods to separate compounds with an enormous structural diversity and only minor differences in chemical or physical properties has to be developed, preferably using only minute quantities.

Although separation and characterization of carbohydrates with HPLC, NMR and mass spectroscopy resulted in major advances within the last decade, a clear need for a highly selective, miniaturized analytical method is evident. Capillary Electrophoresis (CE) has seen a steady growth since its introduction with commercial instrumentation in the late 1980ties, early 1990ties, especially in the life sciences. Automation, miniaturization with the handling of picomole amounts of material, a wide variety of separation modes, that can be easily tailored to a specific analytical problem, and high resolution enable CE as an analytical tool in glycobiology. This book focuses on the various approaches of carbohydrate analysis with CE.

At first, carbohydrate analysis by CE seems inappropriate since sugars lack both charges, a prerequisite for electrophoretic separations, and suitable chromophores or fluorophores for on-column optical detection. However, during the last 6–8 years various strategies have been developed to allow for highly efficient separations and sensitive detection of mono- and polysaccharides, glycoconjugates such as glycoproteins and glycolipids and their constituting glycans. The majority of these strategies are based on carbohydrate specific separation principles such as borate complexation. Derivatization often uses the reducing functional group of the sugar molecule as the point of attack, for example in the reductive amination reaction. Extensive room is provided for a comprehensive overview of CE separation methods to carbohydrate analysis. The enormous number of papers, published until 1998, ranges from monosaccharide determination in fruit juices to the evaluation of the glycan profile in glycoproteins such as immunoglobulines.

Our interest in glycobiology and characterization of carbohydrates was fostered by a Ph.D. thesis co-sponsored by the Corporate Analytical Research of Ciba-Geigy AG in Basel and the Institute of Food Science at the Swiss Federal Institute of Technology ETH Zürich. Both mentors, Prof. Dr. H. Michael Widmer and Prof. Dr. Renato Amadò, strongly supported and guided this program and thus made a unique and fruitful collaboration between industry and academia possible. We would like to dedicate the results of 4 very rewarding research years to Mike Widmer, who had a clear vision to where analytical chemistry would go and the strength to implement a much admired program of analytical research in industry. His guidance in science paired with an unparalleled enthusiasm for new approaches, his ability to create a wonderful work environment for young scientists by picking the right research topics and a team to accomplish bold goals will be greatly missed.

Our thanks also go to the members of the former Corporate Analytical Research group, who we had the good fortune to interact with and who stimulated and encouraged our work. We are indebted especially to Dr. Gerard Bruin and Iris Barmé for their helpful discussion on all aspects of capillary electrophoretic separations and to Catherine Mangin for her contributions to the labeling of complex carbohydrates with novel labels.

Many thanks also to the editor of this book series, Dr. Kevin Altria, for initiating the project and following up with encouragement and very timely and helpful comments on earlier versions of the text. We are grateful to Dr. Angelika Schulz from the Vieweg Verlag for her patience with extended deadlines, her efforts to put all the pieces together and to make sure, style and layout are consistent. Although we made with her help every effort to go to print with an error-free version of the manuscript, we are sure, that as with every scientific publication, some mistakes were overlooked. We take full responsibility for the content and any comments should be directed to us.

Also, we would like to thank the inventors of Internet and email, because without these modern forms of communication, it would not have been possible to co-author a book some 10'000 km and 11 hours flight time apart. Everybody, who has been involved in writing a book knows how much more work is involved than you think when you first get started. Last but not least, we are thankful for the continuos support, encouragement and infinite patience of our families and friends during the long and sometimes difficult times it took to finish this project, which hopefully will be of value for the analytical and glycobiology research community.

Schaffhausen and Hayward, October 1998   Antje Klockow-Beck and Aran Paulus

Table of Contents

List of Abbreviations

ACN	acetonitrile
AGP	α_1-acid glycoprotein
AMAC	aminoacridone
ANSA	aminonaphthalenesulfonic acid
ANDSA	aminonaphthalenedisulfonic acid
2,6-ANS	2-anilinonaphthalene-6-sulfonic acid
ANTS	8-aminonaphthalene-1,3,6-trisulfonic acid
2-AP	2-aminopyridine
APG	N-(4-aminobenzoyl)-L-glutamic acid
API	atmospheric pressure ionization
APTS	8-aminopyrene-3,6,8-trisulfonic acid
6-AQ	6-aminoquinoline
Ara	arabinose
Ar-Ion	Argon-Ion laser
asn	asparagine
BGE	background electrolyte
CAPS	3-cyclohexylamino)-1-propanesulfonic acid
CBQCA	3-(4-carboxybenzoyl)-2-quinoline carboxyaldehyde
CD	cyclodextrin
CDGS	carbohydrate defficient glycoprotein syndrome
CE	capillary electrophoresis
CEC	capillary electrochromatography
CE-MS	capillary electrophoresis – mass spectrometry
CGE	capillary gel electrophoresis
CHAPS	3-((3-cholamidopropyl)-dimethylammonio)propane sulfonic acid
CHES	2-(cyclohexylamino)-ethane sulfonic acid
CHO	chinese hamster ovary
CIEF	capillary isoelectric focusing
CITP	capillary isotachophoresis
CMC	carboxymethyl cellulose
CTAB	cetyltrimethylammonium bromide
CZE	capillary zone electrophoresis
D	diffusion coefficient
DAB	1,4-diaminobutane
DcBr	decamethonium bromide (C_{10}MetBr)
DNA	deoxynucleic acid
DP, dp	degree of polymerization

E	electric field strength
EDTA	ethylenediaminetetraacetic acid
ELISA	enzyme linked immunosorbent assay
EOF	electroosmotic flow
EPO	erythropoietin
ESI	electrospray ionization
ESMS	electrospray mass spectrometry
F	electrostatic force
FAB	fast atom bombardment
FACE	fluorophore assisted carbohydrate analysis
Fru	fructose
Fuc	fucose
GAG	glycosaminoglycans
Gal	galactose
GalA	galacturonic acid
GalNAc	N-acetylgalactosamine
GC	gas chromatography
GCSF	granulocyte-colony-stimulating factor
Glc	glucose
GlcA	glucuronic acid
GlcNAc	N-acetylglucosamine
GPC	gel permeation chromatography
GU	glucose unit
hBMP	human bone morphogenetic protein
hCG	human chorionic gonadotropin
HEC	hydroxyethylcellulose
He-Cd	Helium-Cadmium laser
He-Ne	Helium-Neon laser
Hep	L-glycero-β-D-manno-heptopyranose
hGH	human growth hormone
HPAEC	high performance anion exchange chromatography
HPCE	high performance capillary electrophoresis
HPLC	high performance liquid chromatography
HPMC	hydroxypropylmethylcellulose
hrFVIIa	human recombinant factor VIIa
HSA	human serum albumine
HxBr	hexamethonium bromide (C_6MetBr)
HxCl	hexamethonium chloride (C_6MetCl)
IEF	isoelectric focussing
ID	inner diameter
Ido	idose
IdoA	iduronic acid
IgG	immunoglobulin G

ITP   isotachophoresis

KDO   3-deoxy-β-D-manno-octopyranulosonic acid

L total capillary length
l effective capillary length
LC   liquid chromatography
LE   leading electrolyte
LIF   laser induced fluorescence
LMW   low molecular weight
LOD   limit of detection
LOS   lipooligosaccharides
LPS   lipopolysaccharides

MALDI   matrix assisted laser desorption ionization
Man   mannose
MECC   micellar electrokinetic capillary chromatography
MS   mass spectrometry

N plate number
NBD   7-nitro-2-oxa-1,3diazole
NDA   napthalene-2,3-dicarboxyaldehyde
NeuAc   N-acetylneuraminic acid
NeuGlc   N-glycolylneuraminic acid
NMR   nuclear magnetic resonance spectroscopy

OTLC   open tubular liquid chromatography
OPA   o-phthaldialdehyde

PA   polyacrylamide
PAD   pulsed amperometric detection
PAGE   polyacrylamide gel electrophoresis
PD   plasma desorption
PEG   polyethyleneglycol
PEO   polyethyleneoxide
PMP   1-phenyl-3-methyl-5-pyrazolone
PNGase F   peptide-N-glycosidase F
PVA   polyvinylalcohol

q charge

r radius of a spherical molecule
RAAM   reagent array analysis method
RI   refractive index
Rha   rhamnose
RNAse   ribonuclease
RP-HPLC   reversed phase high performance liquid chromatography
R_S resolution
RSD   relative standard deviation

SA	sulfanilic acid
SDS	sodium dodecylsulfate
TEA	triethylamine
TE	terminating electrolyte
T_f	transferrin
t_M	migration time
TLC	thin layer chromatography
TRSE	5-carboxytetramethylrhodamine succinimidylester
tPA	tissue plasminogen activator
TTAB	tetradecyltrimethylammonium bromide
UV	ultraviolet
V	voltage
v_{eo}	electroosmotic velocity
Xyl	xylose
ε	dielectric constant
μ_{ep}	electrophoretic mobility
μ_{eo}	electroosmotic mobility
η	viscosity
σ^2	peak variance
ζ	zeta-potential

1 Introduction

On a mass scale carbohydrates are the most abundant biomolecules on the face of the earth. Produced by photosynthesis from carbondioxide and water, they serve plants for both their nutritional and physiological needs. However, carbohydrates are not only essential in the plant world, as they play essential roles in a number of biological systems. In particular, carbohydrates are involved in energy storage systems for most forms of life on earth. Being part of the nutritial intake for pro- and eucaryotic organisms such as bacteria, animals and humans, carbohydrates are frequently converted into more complex compounds and serve as the principle source of metabolic energy (starch and glycogen). In plants and bacteria, they act as structural elements in cell walls (cellulose). In insects, they are present in the exoskeleton in the form of chitin, which is also the basis of the shells of crabs and lobsters. Proteoglycans serve as lubricants in joints. Last but not least, carbohydrates are involved in the detoxification and excretion of chemicals in animals and humans by forming water soluble glucuronides.

Carbohydrates are an essential part of human diet, both as soluble mono- and oligomers and as insoluble polymeric fibers. Several mono- and disaccharides, among them saccharose, glucose, fructose, maltose and lactose, serve as sweeteners. Polysaccharides such as carrageenan and pectin as well as modified starches and celluloses are used as gelling and thickening agents in food, as stabilizers for emulsions or as modifiers of food textures. Finally, the color of cola drinks and the flavor of baked bread are due to the Maillard browning of amino acids and carbohydrates.

As integral components of larger biopolymers such as glycoproteins (see Figure 1.1), glycolipids and proteoglycans, carbohydrates affect their intramolecular and intermolecular functions. For example, glycosylation influences the folding of proteins, thereby influencing proteolysis rate, thermal stability and solubility. Attached to cell surface proteins, oligosaccharides allow cells to communicate with each other through very specific and at this time not well understood intercellular recognition and adhesion. Glycosylation affects the translocation of the glycoprotein from one cellular compartment to another as well as its clearance from the body. Glycoproteins also serve as receptors for antibodies and hormones and are known to play an important role in several pathological processes such as inflammation, infection or oncology.

Glycosylation of proteins is prevalent in mammalian cells. Consequently, the second generation of therapeutic proteins produced by recombinant techniques with mammalian cell lines are also often glycosylated. The cDNA sequence reveals no information on the degree of glycosylation or the carbohydrate sequences. Only the position of potential sites for N-glycosylation is apparent. Therefore, the information whether a potential N-glycosylation site is actually glycosylated or not and if a recombinant glycoprotein is glycosylated at serine or threonine residues (O-glycosylation) can only be obtained experimentally [1]. The degree of glycosylation and the structure and composition of the glycans, depend on the biosynthetic conditions and the cell line in which the protein is expressed. Additionally, the different

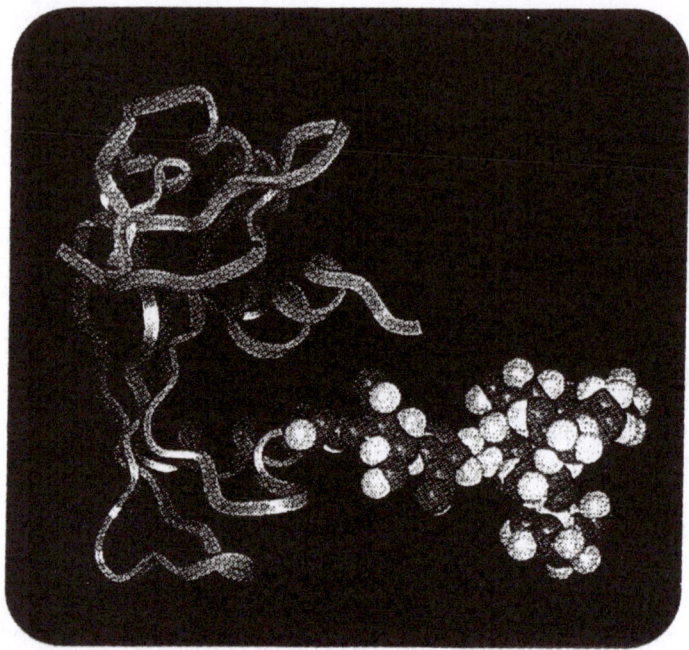

Figure 1.1   Molecular model of mannose-9 on ribonuclease B (from [5], with permission)

glycosylation sites within a glycoprotein can have different populations of an attached carbohydrate. Control of this diversity in glycosylation is of increasing interest and importance in biotechnology, since it affects biological activity, pharmacokinetics, cellular uptake and specificity of recombinant glycoproteins.

The enormous molecular diversity of oligo- and polysaccharides poses a fabulous analytical problem. For a complete structural elucidation, a highly selective and efficient separation technique must be coupled with an information-rich structural analysis method. Among the separation methods used to differentiate closely related carbohydrate solutes, are thin layer chromatography (TLC), gas chromatography (GC), high performance liquid chromatography (HPLC) and electrophoresis on slab gels. To elucidate the structure, various spectroscopic methods such as nuclear magnetic resonance spectroscopy (NMR) and mass spectrometry (MS) have been used.

The analytical problems for carbohydrate elucidation and characterization range from separation and determination of monosaccharides to such challenging tasks as assessing glycoprotein microheterogeneity, oligosaccharide mapping or the characterization of polysaccharide polydispersity. The enormous number of possible carbohydrate isomers, due to variable linkage forms and the α/β-anomerity of the glycosidic bonds requires extreme selectivity. The polar and nonvolatile nature of carbohydrates and the absence of chromophoric and fluorophoric functional groups, add to the analytical difficulties with carbohydrates.

The analysis of polymeric carbohydrates, especially of complex glycoprotein glycans comprises the following steps:

1. monosaccharide composition analysis
2. profile analysis
3. structure analysis

Monosaccharide composition analysis is required to determine the specific monosaccharides present in glycans and their relative abundance. Acid hydrolysis of the carbohydrate polymers followed by GC allows for the determination of the absolute and relative molar ratios of the monosaccharide residues. The *profile analysis* serves to determine the number and types of glycans bound to a glycoprotein and their relative molar contribution. The types of glycan profiling commonly used are mass profiling i. e. by matrix assisted laser desorption ionization-time-of-flight mass spectroscopy (MALDI-TOF-MS), size profiling by gel permeation chromatography (GPC) and charge profiling by using anion exchange chromatography. To elucidate the primary structure of polymeric carbohydrates, a *structure analysis* is required comprising the determination of

1. the nature, order and ring conformation of individual monosaccharides,
2. the anomericity (α- or β-linkage) of individual glycosidic bonds,
3. substitution patterns and branching points,
4. the absolute stereochemistry of individual residues (D- or L-),
5. the nature and location of chemical substituents (e.g. sulfate, phosphate, O-methyl) on a given monosaccharide.

Since no single technique is able to provide all this information, exact structural characterization relies on the combined use of several physical, chemical and biochemical techniques such as NMR, MS and Enzymatic Analysis.

^{1}H-NMR is potentially the most informative technique used in carbohydrate structure elucidation. NMR is totally non destructive but requires in contrast to other characterization techniques large amounts of purified material (see Table 1.1). Structural analysis of unknown carbohydrates can be carried out either by *fingerprint matching*, comparing the spectrum of an unknown structure with the spectrum of known structures stored in large databases, or by *sequencing* of oligosaccharides. The latter approach requires the combined use of several ^{1}H-NMR and ^{13}C-NMR techniques to identify the monosaccharides and to determine their anomericity and linkage (see Figure 1.2 [2]).

Table 1.1 Methods for glycan structure analysis

	Minimum sample requirement	Monosaccharide composition	Anomericity and linkage	Derivatization	Operator skill level
NMR	50 nmoles	yes	yes	no	very high
MS					
ESI	low pmoles	Hex/HexNAc	no	yes	high
MALDI	low pmoles	Hex/HexNAc	no	no	low
FAB	10 pmoles	Hex/HexNAc	limited	yes	high
Enzymatic Sequencing	150 pmoles	yes	yes	yes	low

Hex = hexose, HexNAc = N-acetylhexosamine

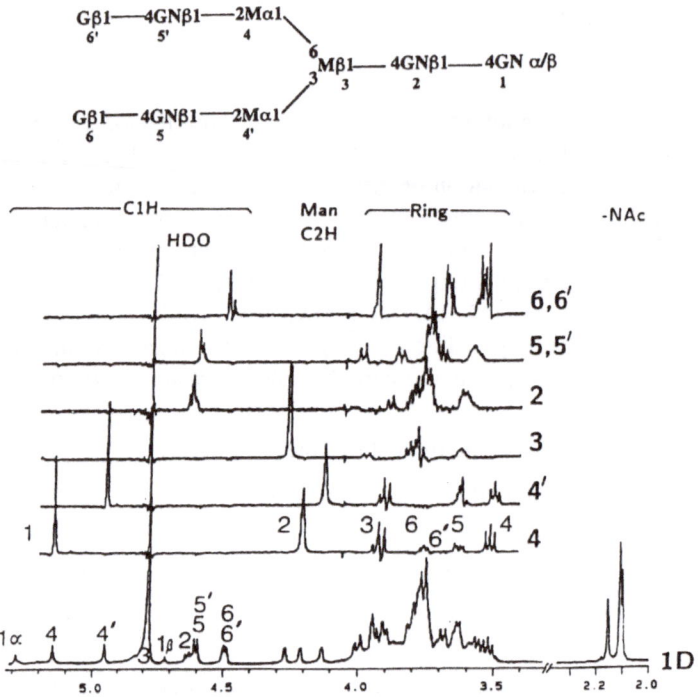

Figure 1.2   One-dimensional ^{1}H-NMR spectrum (*bottom*) of a complex oligosaccharide (asialo biantennary, shown above) and the subspectrum for each individual monosaccharide ring.
The C1H (anomeric), mannose C2H, ring, and N-acetyl regions of the 1D spectrum are
shown. The residue assignments for the anomeric resonances are given on the 1D spectrum and on the right-hand side for the subspectrum (residue numbering sequence is give
on the structure). Each separate resonance is also assigned in the subspectrum of residue
4 (from [2], with permission).

MS is a highly sensitive technique requiring only picomol or femtomol quantities of material (see Table 1.1). The phenomenal success of MS techniques in the last decade arises
from the wealth of structural information of the analytes and the possibility to couple it to
powerful separation techniques such as GC and HPLC. The first MS technique applied to
carbohydrate analysis was *Electron Impact Ionization (EI)*. It is still used in combination
with GC for methylation analysis [2]. The introduction of soft ionization techniques such as
Fast Atom Bombardment (FAB) represented a major breakthrough for carbohydrate analysis
by MS. FAB often yields molecular ions in addition to fragmentation providing valuable sequence and branching information [2]. With the introduction of *Electrospray Ionization
(ESI)* and *MALDI,* the mass range of compounds to be examined by MS techniques was tremendously increased. Both ESI-MS and MALDI-MS are capable of molecular weight determination in excess of 200 kDa with a mass accuracy of ± 0.01% and ± 0.1%, respectively
[3]. In ESI, multiple charged fragments are generated, allowing for the accurate mass determination of glycoproteins and glycopeptides. Due to its high resolution (± 2000) ESI-MS

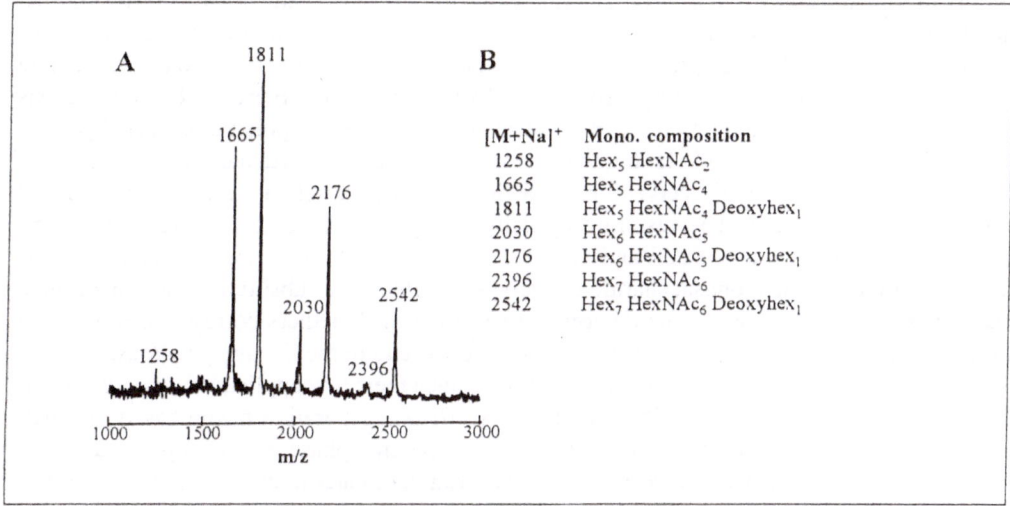

Figure 1.3 MALDI-MS analysis of desialylated N-glycans released from recombinant human interferon-γ by PNGase F. (A) Mass spectrum: N-glycans are [M*Na]$^+$ ions; m/z values are indicated. (B) Monosaccharide compositions were assigned using (CCSD). Hex = hexose; HexNAc = N-acetylhexosamine; Deoxyhex = deoxyhexose (from [3], with permission).

can also be applied to glycoform analysis. In contrast to ESI-MS, MALDI is particularly simple and quick to use. The soft laser desorption of the sample produces predominantly molecular ions and has been used to measure the molecular masses of low picomol amounts of glycoproteins and oligosaccharides (see Figure 1.3) without the need for derivatization [4].

A slightly more complex yet still elegant way to elucidate the primary structure of a glycan is enzymatic analysis using specific exo- and endoglycosidases. Enzymes are commercially available to cleave a monosaccharide with a specificity which also allows to deduce the anomericity of the glycosidic linkage and the absolute D/L configuration. Usually, enzymatic analysis of oligosaccharide structures is performed sequentially by exposing the oligosaccharide in solution to glycosidases. After each incubation step the oligosaccharide product is identified and the loss of monosaccharide is measured. A new process for primary sequence analysis uses the glycosidases not sequentially, but in multiple defined mixtures. In the so called Reagent Array Analysis Method or RAAM [2, 5], several aliquots of a glycan are incubated with a precisely defined mixture of exoglycosidases which fragment the glycan until a linkage is reached that is resistant to all the enzymes present in the mixture. By omitting one or more different exoglycosidases from each mixture, different 'stop point' fragments are generated. Chromatographic separation of the combined 'stop point' fragments generates a pattern representing a 'signature' of the specific oligosaccharide.

Capillary GC was the first analytical separation technique allowing for the accurate quantitation of carbohydrates. However, in order to apply GC to carbohydrate analysis, a derivatization step is necessary. Generally, two methods are used to generate volatile carbohydrate derivatives, trimethylsilylation [6] and alditol peracetylation [7]. While trimethylsilylation results in a mixture of trimethylsilyl derivatives of the pyranose and furanose forms, the α- and β-anomers and sometimes even the open chain form of each monosaccharide with consequently complex chromatographic patterns, alditol peracetylation forms only one single peak per sugar due to ring opening through borohydride reduction. For composition analysis and quantitative characterization of monosaccharide linkages in polymeric carbohydrates such as polysaccharides, methylation analysis, first introduced by Hakamori [8], is used in combination with alditol peracetylation. In a first step the carbohydrate is methylated and subsequently hydrolyzed. The resulting partially methylated monomers are reduced to the corresponding alditols, acetylated and detected on-line by MS. This way, an unambiguous identification of all the monosaccharide isomers is possible. The substitution pattern within a carbohydrate polymer can also be determined since the methylethers correspond to the free hydroxy groups of the monosaccharides and the acetylesters to the linkage positions.

HPLC methods have seen a rapid development since their first application to carbohydrate analysis in the mid 1970ties. The use of highly efficient separation media based on small silica particles contributed to the increasing success. Normal phase, reversed phase and size exclusion systems have been described for the separation of carbohydrates [9]. Native carbohydrates are no suitable solutes for common HPLC analysis for two reasons. First, detection is a problem since their UV absorbance is generally very low, even at a wavelength of 200 nm. In addition, they are so hydrophilic, that they are barely retained on the popular reversed phase columns resulting in a loss of resolution. A derivatization with labels such as 2-aminopyridine, dansylhydrazine, fluorenylmethyl chloroformate (Fmoc) or 1-phenyl-3-methyl-5-pyrazolone did improve both selectivity and detection somewhat, but failed to meet the stringent requirements in glycobiology.

This situation only changed in the second half of the 1980ties with the introduction of High Performance Anion Exchange Chromatography (HPAEC) with Pulsed Amperometric Detection (PAD). The development of strong polymeric ion exchange stationary phases, which unlike silica based stationary phases are stable even at the extreme of pH 14 and the pulsing of the amperometric detector, which avoided electrode fouling by the carbohydrate oxidation products, revolutionized carbohydrate analysis. The most significant improvement of HPAEC is its ability to resolve neutral and anionic positional isomers. Even two glycoprotein derived oligosaccharides differing only in the linkage position of a single sugar residue could be separated. As demonstrated in Figure 1.4, fractionation of sialylated oligosaccharides from bovine fetuin was achieved based on their degree of sialyzation. The fine-structure in the sialylated groups is due to isomers which differ in composition and/or linkage of the non-charged portion of the glycan [10, 11]. Inherent shortcomings of HPAEC-PAD are the strong alkaline medium that could possibly induce epimerization and degradation of reducing carbohydrates and it is non-discriminative character in the presence of amino acids and peptides [12].

GPC, in use for more than a decade in oligo- and polysaccharide analysis, allows the determination of glucose equivalents in a recently introduced automatic *Glycosequencer* (Oxford Glycosystems). The oligosaccharide sequence and composition can be deduced from the glucose equivalents, based on the hydrodynamic volumes of the oligosaccharides, after a sequential digestion with specific exoglycosidases.

Parallel to the developments in chromatography, another highly selective separation method, electrophoresis, was applied to carbohydrate analysis. Electrophoretic separations have

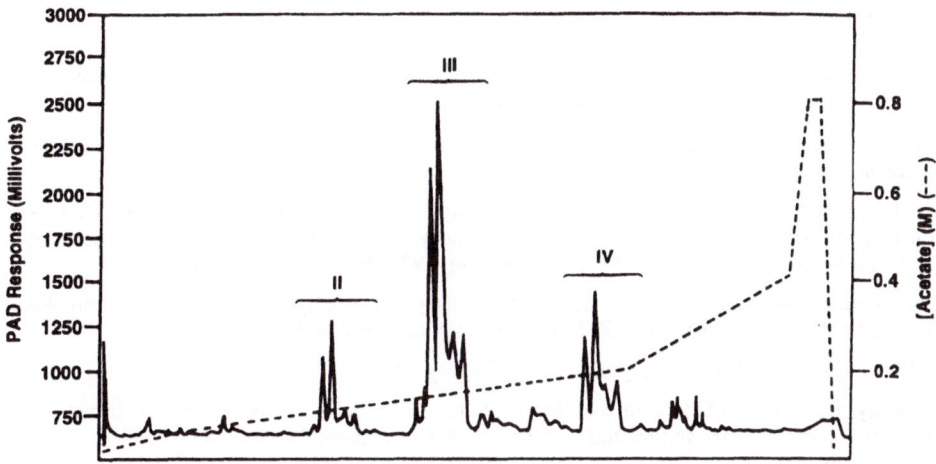

Figure 1.4 Fingerprinting of oligosaccharides on a Dionex HPAE-PAD II system at high pH. Asn-linked oligosaccharides released from bovine fetuin. The clusters of peaks denoted by *n*, *II*, *III* and *IV* represent neutral, di-sialylated, tri-sialylated and tetra-sialylated oligosaccharides, respectively (from [10], with permission).

been demonstrated for polysaccharides and their degradation products on different supports such as paper, cellulose acetates [13], glass fiber paper [14, 15] and reversed phase silica gel [14]. However, slab gel electrophoresis, the method of choice for the separation of proteins and nucleic acids, has been of only limited use. Recently, Jackson [16, 17] introduced a novel strategy based on reductive amination of carbohydrates with a suitable fluorophore, such as aminonaphthalenetrisulfonic acid (ANTS) and 2-aminoacridone prior to slab gel electrophoresis. This method is called fluorophore-assisted carbohydrate analysis (FACE). Especially, ANTS labeled saccharides exhibit high resolution due to the high electrophoretic mobilities of the ANTS-conjugates and the combined use of stacking buffer systems. This system was found to be particularly useful for profiling N-linked oligosaccharides after their enzymatic release from glycoproteins.

Unfortunately, slab gel electrophoretic separations are labor intensive manual techniques. Modern analytical methods are characterized by a high degree of automation and increasing miniaturization. The introduction of capillary electrophoresis (CE) by Jorgenson [18, 19] and its subsequent commercialization gave rise to the development of a new tool for carbohydrate analysis. CE is compatible with the minute amounts of sample available in the life sciences. In addition, the ease of quantitation after on-line detection, the high performance with several hundred thousand plates generated in minutes and the aqueous separation media present advantages over alternative separation schemes. At the same time, the increased interest in glycobiology with its need for effective analytical methods to separate and characterize glycoproteins as the first step to understand their role on the molecular level of life depends on improved tools.

This monograph summarizes the development of CE into a method for carbohydrate separation by reviewing and critically evaluating the various approaches described in the literature.

References

1. M.W. Spellman, "Carbohydrate characterization of recombinant glycoproteins of pharmaceutical interest", *Anal. Chem.*, *62* **1990** 1714-1722.

2. R.A. Dwek, C.J. Edge, D.J. Harvey, M.R. Wormald, and R.B. Parekh, "Analysis of glycoprotein-associated oligosaccharides", *Annu. Rev. Biochem.*, *62* **1993** 65-100.

3. D.C. James, and N. Jenkins, "Analysis of N-Glycans by Matrix-Assisted Laser Desorption/Ionization Mass Spectrometry", in "A Laboratory Guide to Glycoconjugate Analysis". (ed. P. Jackson and J.T. Gallagher), Birkhäuser Verlag, **1997**, Basel.

4. C.A. Settineri, and A.L. Burlingame, "Mass Spectrometry of Carbohydrates and Glycoconjugates", in "Carbohydrate Analysis". (ed. Z. El Rassi), Elsevier, **1995**, Amsterdam.

5. Oxford Glycosystems, "Tools for glycobiology", *Catalog*, **1994** .

6. C.C. Sweely, R. Bentley, M. Makita, and W.W. Wells, "Gas-liquid chromatography of trimethylsilyl derivatives of sugars and related substances", *J. Chem. Soc.*, *85* **1963** 2497-2507.

7. A.B. Blakeney, P.J. Harris, R.J. Henry, and B.A. Stone, "A simple and rapid preparation of alditol acetates for monosaccharide analysis", *Carbohydr. Res.*, *113* **1983** 291-299.

8. S.I. Hakamori, "A rapid permethylation of glycolipid, and polysaccharide catalyzed by methylsulfinyl carbanion in dimethyl sulfoxide", *J. Biochem.*, *55* **1964** 205-208.

9. Z. El Rassi (ed.), "Carbohydrate Analysis", Journal of Chromatography Library, Vol. 58, Elsevier, **1995**, Amsterdam.

10. S.B. Yan, Y.B. Chao, and H. van Halbeek, "Novel Asn-linked oligosaccharides terminating in GalNAcβ(1-4)[Fucα(1-3)]GlcNAcβ(1-•) are present in recombinant human Protein C expressed in human kidney 293 cells" *Glycobiology*, *3* **1993** 597.

11. R.R. Townsend, "Analysis of Glycoconjugates using high-pH Anion-Exchange Chromatography", in "Carbohydrate Analysis". (ed. Z. El Rassi), Elsevier, **1995**, Amsterdam.

12. Y.C. Lee, "High-performance anion-exchange chromatography for carbohydrate analysis" *Anal. Biochem.*, *189* **1990** 151-162.

13. H. Weigel, "Paper Electrophoresis of Carbohydrates" *Adv. Carbohydr. Chem.*, *18* **1963** 61.

14. H. Scherz, "Thin-layer electrophoretic separation of monosaccharides, oligosaccharides and related compounds on reverse phase silica gel" *Electrophoresis*, *11* **1990** 18-22.

15. B. Bettler, R. Amado, and H. Neukom, "Electrophoretic separation of sugars and hydrolyzates of polysaccharides on silylated glass fiber paper" *J. Chromatogr.*, *498* **1990** 213-221.

16. P. Jackson, "The use of polyacrylamide-gel electrophoresis for the high-resolution separation of reducing saccharides labelled with the fluorophore 8-aminonaphthalene-1,3,6-trisulfonic acid", *Biochem. J.*, *270* **1990** 705-713.

17. P. Jackson, "Polyacrylamide gel electrophoresis of reducing saccharides labeled with the fluorophore 2-aminoacridone: subpicomolar detection using an imaging system based on a cooled charge-coupled device", *Anal. Biochem.*, 196 **1991** 238-244.

18. J.W. Jorgenson, and K.D. Lukacs, "Zone electrophoresis in open-tubular glass capillaries", *Anal. Chem.*, *53* **1981** 1298-1302.

19. J.W. Jorgenson, and K.D. Lukacs, "High-resolution separations based on electrophoresis and electroosmosis", *J. Chromatogr.*, *218* **1981** 209-216.

2 Capillary electrophoresis, instrumentation and modes

The success of open tube capillaries over packed columns in gas chromatography (GC) initiated the development of open tubular liquid chromatography (OTLC) in the late 70ties, in an attempt to further increase the performance of high performance liquid chromatography (HPLC) [1]. Theoretical analysis of the chromatographic process clearly showed, that miniaturization of the separation column dimensions will lead to decreased band broadening and therefore to better and/or faster analysis. However, diffusion constants are 2 to 3 orders of magnitudes smaller in liquids compared to gaseous system. In combination with the Poiseuille flow profile of the pressure generated flow, which results in fast longitudinal transport in the middle of the column and slow transport near the column wall, this leads to a relatively large diffusion controlled contribution to band broadening. Consequently, liquid separation systems did not show the same performance increase when switching from packed columns to open column operation compared to GC systems. Theoretical calculations indicated that only with extremely small capillaries of 10 μm inner diameter (ID) or less open capillary operation in liquids could outperform packed columns.

However, working with capillaries this small represented a tremendous challenge for analytical instrumentation. A 10 μm ID capillary of 1 m has a total volume of less than 80 nanoliter. In order to operate these capillaries, all relevant chromatographic parameters had to be adjusted. For example, pumping systems to generate flow rates in the order of nanoliters per minute had to be developed as well as injection schemes for picoliter injection volumes. Since the detection volume also decreases to the picoliter range, miniaturization of UV/Vis detection as the workhorse detection system in traditional HPLC was not suitable. More sensitive detection methods such as electrochemical detection or fluorescence detection to measure sub-nanomolar concentrations were at the time not sufficiently developed to allow their routine application to capillary LC systems.

In summary, a miniaturization of almost 3 orders of magnitude from the traditional 4.6 mm ID packed HPLC column to 10 μm ID or less capillaries proofed not to be feasible at the time because of numerous technical and instrumental problems. If, however, capillaries of around 100 μm ID could be used, the technical hurdles were less daunting. Since the separation performance is directly limited by the Poiseuille flow profile and diffusion in the liquid system, an alternative flow generation could be a solution. Pretorius et al. had shown as early as 1974, that rather than using a pump for flow generation, electroosmosis could drive a liquid through a packed bed [2]. The main advantage of electroosmotic flow (EOF) is its plug flow profile (Figure 2.1), which limits axial diffusion.

In 1981, Jorgenson used EOF to generate a flow in a fused silica capillary of 75 μm ID [3]. Negatively charged labeled peptide fragments could be separated under alkaline buffer conditions. This work is generally seen as an early milestone in the development of current CE as a liquid based separation method. The electroosmotic flow rates are in the same order

Figure 2.1 Flow profile in an electrically driven system

of magnitude as linear flow rates in HPLC. Due to the flat flow profile theoretical plate numbers of several 100.000 plates per column could be achieved using capillaries up to 200 µm ID. Thus the plug flow profile has been the most important reason for the superior performance of CE versus HPLC or OTLC.

In this chapter, some basic electrophoretic parameters such as electrophoretic mobility and resolution will be explained. Also, the EOF as the fundamental driving force in CE is discussed. A short overview of the components of a CE instrument followed by a description of the different CE modes and their general field of applications concludes this chapter.

2.1 Electrophoretic mobility

In a constant electric field, ions experience an electrostatic force F, proportional to the electric field strength (E) and the charge (q) of the ion.

$$F = q E \tag{2.1}$$

This electrostatic force causes a migration of the charged species toward the opposite charged electrode with a constant velocity, v_{ep}, again being proportional to the applied electric field. The proportionality factor is called electrophoretic mobility, μ_{ep}

$$v_{ep} = \mu_{ep} E = \mu_{ep} V/L \tag{2.2}$$

with V the applied voltage and L the total length of the capillary. The electrophoretic mobility is a characteristic property of a given species in a defined medium at a defined temperature. According to Stokes' Law, μ_{ep} can be expressed as

$$\mu_{ep} = q/6 \, \pi\eta r \tag{2.3}$$

with η being the viscosity of the electrolyte and r the radius of a spherical molecule. It follows from equation (2.3) that the migration of a species in an electric field depends on its charge, size and shape, as well as on the viscous drag of the solvent. Since viscosity is a function of temperature, electrophoretic mobility increases roughly 2% per °C with temperature [4, 5].

2.2 Fused silica surface and electroosmotic flow

The basic transport mechanism in open tube CE is the EOF [6]. Since CE is practiced almost exclusively in fused silica capillaries, EOF will here only be discussed with reference to silica surface characteristics. Chemically, the surface of an uncoated fused silica capillary is covered with silanol groups (Figure 2.2). At pH > 3 silanol groups can dissociate generating an excess of negative charges at the capillary surface. In an electrical field, cations from the buffer electrolyte are attracted towards the negatively charged silanol groups, forming an ionic layer next to the capillary wall. The charge density of this layer decreases exponentially with increasing distance from the capillary wall. Upon the application of an electrical field, the cations in the double layer next to the wall start moving towards the cathode. However, since the cations in an aqueous solution are surrounded by a cloud of solvating water molecules, this hydration shell also moves towards the cathode, resulting in a mass flow from the anode to the cathode.

The electroosmotic velocity is defined as

$$v_{eo} = \mu_{eo} E \qquad (2.4)$$

and the electroosmotic mobility, μ_{eo}, as

$$\mu_{eo} = \varepsilon \, \zeta / 4\pi\eta \qquad (2.5)$$

In chromatographic separations, reproducible flow conditions of the mobile phase and thus reproducible retention times are achieved through a precise pressure control. This is in contrast to CE, where the electrophoretic velocity not only depends on the applied electrical field but also and to a greater extent on chemical and physical parameters of the capillary wall and the buffer medium. Unfortunately, those parameters are not as easily and independently manipulated as the pressure in a pressure driven flow system. Nevertheless, reproducible EOF conditions are key to successful CE separations.

As expressed in equation 2.5, the EOF is a function of the dielectric constant ε, the zeta-potential ζ and the viscosity η of the separation medium. The zeta-potential is a function of the surface charge density on the capillary wall, and is proportional to the double layer thickness and inversely proportional to the ionic strength of the buffer medium.

Surface charge density in fused silica capillaries strongly depends on the pH. With an estimated pK of the silanol groups between 3 and 5, surface charge density is low at acidic

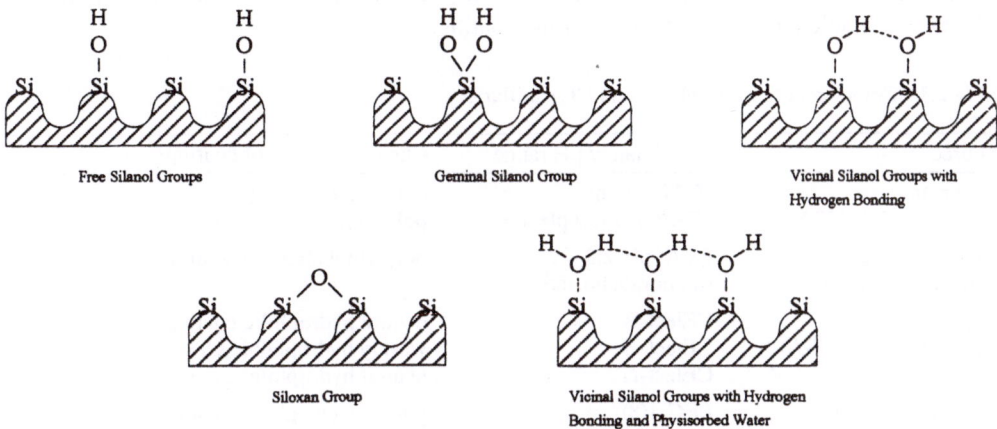

Figure 2.2 Silanol groups on a fused silica surface

pH and increases with increasing pH. Above pH 9, all silanol groups are assumed to be ionized, and therefore, surface charge density is maximal under basic conditions.

Using an electrical field strength of 300 V/cm electroosmotic velocities of up to 4 mm/sec are observed in untreated fused silica capillaries. These EOF values are considerably higher than the electrophoretic mobilities of most analytes of interest. The net mobility of the analytes is the vectorial sum of both their electrophoretic mobility and the electroosmotic velocity

$$\mu_{net} = \mu_{ep} + \mu_{eo} \tag{2.6}$$

With a large EOF, the net mobility will be determined by the magnitude and direction of the EOF. This fact has important consequences for on-line detection. Since at higher pH's (pH > 5) the EOF is the dominant flow factor and for fused silica capillaries in direction from anode to cathode, the online detector has to be placed at the cathodic side of the capillary, independent of the analyte charge. Both negative and positive analytes will be swept towards the detector.

At low pH (pH 2–3) the silanol groups of fused silica are essentially not ionized, consequently the zeta-potential at the capillary wall to buffer interface is very low as is the resulting EOF. In fact, at this acidic pH the EOF is negligible compared to the electrophoretic mobility of most charged compounds. In this case, the magnitude and direction of the electrophoretic mobility of the analytes of interest determine their net mobility. Positive and negative analytes can no longer be separated together in one run. In order to detect negatively charged analytes, the polarity of the electrical field should be reversed compared to conditions with large EOF, injection occurs at the cathodic side and the detection at the anodic side.

Another way to manipulate the surface charge density and thus the magnitude of the EOF is a chemical modification of the capillary surface. Both neutral and positive coatings have been developed. The chemistries of these coatings are also tailored to prevent the absorption of positively charged components such as proteins to the negatively charged fused silica surface. A number of commercial companies offer coated capillaries for various applications. For most applications, a stable hydrophilic coating, suitable for a wide pH range, is desirable. Table 2.1 lists a selection of commercial coatings including the recommended pH range and when available the chemical nature of the coating.

Table 2.1 Commercially available coated CE capillaries

Source	Brand name/ pH range	Chemical nature of coating
Beckman (Fullerton, CA, USA)	eCAP Amine eCAP neutral/ pH 4-8	polyamine coating polyacrylamide coating
Hewlett Packard (Waldbronn Germany)	PVA/ pH 2.5 – 9 (no borate buffers)	poly(vinyl alcohol) coating
Supelco, Inc. (Bellafonte, PA, USA)	CElect-P CElect-H CElect-N	neutral hydrophilic coating weakly hydrophobic C_1 coating neutral hydrophilic coating
Scientific Resources (Eatontown, NJ, USA)	PEG-100	polyethylene glycol coating

Table 2.2 EOF in different electrolyte systems

Electrolyte system	Concentration [mM]	pH	EOF [$10^{-5}cm^2/Vs$]
phosphate	10	2.5	4.2
phosphate	50	2.5	1.2
phosphate	100	2.5	< 0.7
citrate	50	2.5	2.6
phosphate	50	9.0	49
borate	50	9.0	58
borate	150	9.5	46

The EOF is also inversely proportional to the viscosity of the buffer solution. Therefore the EOF can be manipulated through viscous buffer components such as ethylene glycol, iso-propanol and other hydrophilic additives. This may be useful in cases, where the relative mobility differences of two solutes ($\Delta\mu_{ep}$) is small compared to the EOF.

Last but not least, the nature of the buffer and the buffer concentration influences EOF by changing the thickness of the double layer. Table 2.2 lists some acidic and basic electrolyte compositions and their corresponding EOF [7].

2.3 Plate number, migration time and resolution

The plate number in separation systems is defined as [8]

$$N = L^2/\sigma^2 \tag{2.7}$$

with σ^2 being the variance in length units. Assuming longitudinal diffusion as the only source of band broadening, the variance σ^2 is defined as

$$\sigma^2 = 2\,D\,t_M \tag{2.8}$$

with D the diffusion coefficient and t_M the migration time of a solute of interest. t_M, obtained experimentally is

$$t_M = l/v_{ep} \tag{2.9}$$

and by substituting v_{ep} according to equation 2.2 becomes

$$t_M = L\,l/\mu_{net}\,V \tag{2.10}$$

with l the effective capillary length from injection to detection. Substituting equation 2.10 in 2.8 yields

$$\sigma^2 = 2\,D\,L^2/\mu_{net}\,V \tag{2.11}$$

and equation 2.11 into equation 2.7 results in

$$N = \mu_{net}\,V/2\,D \tag{2.12}$$

It is interesting to note, that the plate number is independent of the capillary length. Also, increasing the voltage will result in higher plate counts as long as Joule heating does not add to additional band broadening. Fast solutes will separate with high efficiency because the time for diffusion in the system is short. Therefore highly charged small solutes will move in narrow bands as will solutes with low diffusion coefficients, such as proteins or DNA.

The resolution R_S of two analyte zones can be expressed according to Giddings as

$$R_S = \frac{1}{4}\sqrt{N\frac{\Delta\mu_{ep}}{\overline{\mu}_{ep}}} \qquad (2.13)$$

with $\Delta\mu_{ep}$ the mobility difference of the two solutes and $\overline{\mu}_{ep}$ the average mobility. In separation systems with EOF, equation 2.14 changes to

$$R_S = \frac{1}{4}\sqrt{N\frac{\Delta\mu_{ep}}{\mu_{eo}+\overline{\mu}_{ep}}} \qquad (2.14)$$

Equation 2.14 consists of an efficiency term $\frac{1}{4}\sqrt{N}$ and a selectivity term $\Delta\mu_{ep}/\left(\mu_{eo}+\overline{\mu}_{ep}\right)$.

To increase the resolution, an increase of voltage, which translates in an increased plate count, is not very efficient since both scale only by the square root. However, a selectivity increase has a more pronounced effect. For selectivity, the ratio of the absolute value of μ_{ep} to μ_{eo} will obviously strongly affect the resolution. With the EOF in the direction of solute migration, resolution will decrease and vice versa. Therefore, maximum resolution will be obtained when the EOF has the same value as μ_{ep} but opposite direction ($\mu_{eo} = -\mu_{ep}$). However, the price is infinite analysis time.

2.4 Instrumentation

2.4.1 Power supply and cooling system

A scheme of a CE instrument is depicted in Figure 2.3. A working CE set-up includes the separation capillary, a power supply, a detection system, a cooling system and a data system [9]. Compared to a HPLC system, the main difference in CE instrumentation is the exchange of the solvent pumps by a power supply. Current CE instrumentation uses power supplies capable of delivering ±30 kV with upper current limits of 200–400 μAs. Total power consumption is limited to 12 W, but in practical separation rarely supersedes 1 W. With a typical capillary length of 25–50 cm, electrical field strengths of up to 1000 V/cm are possible. This compares to 10–50 V/cm in classical slab gel electrophoresis, a 20–100 fold increase. Since in electrophoretic separations, speed, performance and resolution are directly coupled to the applied field, this demonstrates the superior performance of CE over slab gels.

In order to prevent the Joule heat to built up within the capillary, thus increasing the current and resulting in shifting migration times, cooling is necessary to remove excess heat from the capillary. Commercial instrumentation uses flowing streams of temperature, controlled liquids or air [10, 11].

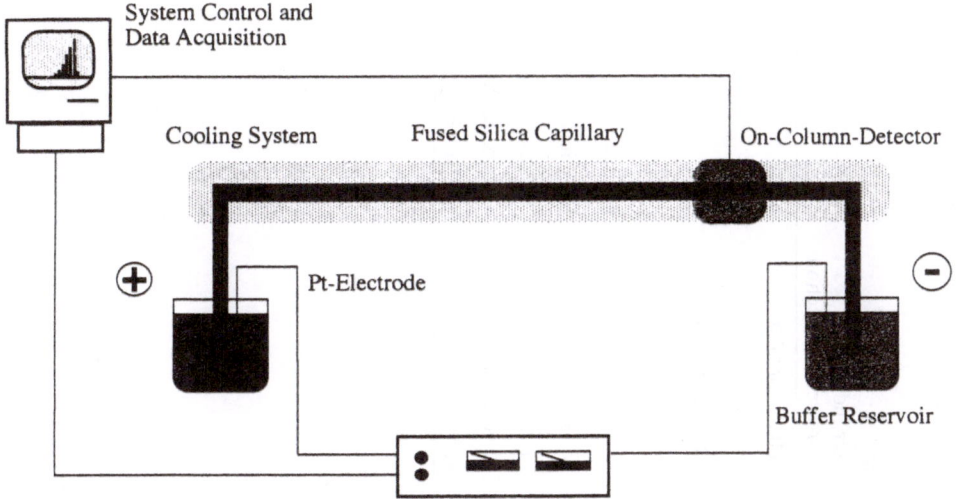

System Control and
Data Acquisition

Cooling System   Fused Silica Capillary   On-Column-Detector

Pt-Electrode

Buffer Reservoir

High Voltage Power Supply

Figure 2.3 Schematic CE instrumentation

2.4.2 Capillaries

The capillaries in CE are made almost exclusively from fused silica. Only a few isolated reports use plastic capillaries, most notably Teflon capillaries. Fused silica capillaries can be made available in almost any inner diameter between 1 and 530 µm. For protection, the fused silica capillaries are coated on their outside with a 10–20 µm thick film of polyimide making them more flexible and therefore easier to handle. Today, polyimide coated fused silica capillaries with inner diameters of 50–100 µm and outer diameters around 360 µm are most common in commercial instrumentation.

It is worthwhile to compare the volumes of a 50 cm capillary of various diameter, which are typically used in CE (Table 2.3). The overall volume ranges from 40.0 nL for a 10 µm ID capillary to 15.7 µl for a 200 µm ID capillary. Since only about 1% of the column volume should be used for injection and detection, these volumes are in the order of 0.4–160 nL. This puts a heavy burden on sensitive detection methods for CE.

2.4.3 Detection

2.4.3.1 UV detection

Fused silica is optically pure quartz glass and by burning off a small section of the polyimide coating, a perfect detection window for spectroscopic detection methods is obtained. UV detection has been a very popular detection method for liquid separation systems because of

Table 2.3 Capillary volumes, injection volumes and mass detection limits in CE

Capillary ID [μm]	Capillary Volume	Injection Volume	Mass detection limits			
			10^{-6} M UV detection	10^{-8} M Electrochemical or Fluorescence Detection	10^{-11} M Laser Induced Fluorescence Detection	
200 μm	15.7 μL	160 nL	160 femtomoles	160 attomoles	1600 zeptomoles	
100 μm	3.9 μL	40 nL	40 femtomoles	40 attomoles	400 zeptomoles	
75 μm	2.2 μL	20 nL	20 femtomoles	20 attomoles	200 zeptomoles	
50 μm	1.0 μL	10 nL	10 femtomoles	10 attomoles	100 zeptomoles	
25 μm	245.0 nL	2450 pL	2500 attomoles	25 attomoles	25 zeptomoles	2400 molecules
10 μm	40.0 nL	400 pL	400 attomoles	4 attomoles	4 zeptomoles	600 molecules
5 μm	10.0 nL	100 pL	100 attomoles	1 attomole	1 zeptomole	90 molecules
2 μm	1.5 nL	15 pL	15 attomoles	150 zeptomoles	150 yoctomoles	20 molecules
1 μm	0.4 nL	4 pL	4 attomoles	40 zeptomoles	40 yoctomoles	

Capillary length: 50 cm
Injection Plug Length: 0.5 cm
ID = inner diameter

Table 2.4 Detection methods and detection limits in CE

Detection Method	Detection Limit [mol/L]	Reference
UV-absorbance	10^{-5}–10^{-6}	[15–18]
indirect UV	10^{-5}	[19–21]
LIF	10^{-12}	[22, 23]
indirect LIF	10^{-7}	[24–26]
conductivity	10^{-6}	[27, 28]
amperometry	10^{-8}	[29, 30]
termo optical absorbance	10^{-8}	[31, 32]
mass spectrometry	10^{-7}–10^{-8}	[33, 34]

its almost universal suitability. For its use in CE instrumentation commercial HPLC UV detectors with deuterium lamps were modified for on-column detection and became the general detector for almost all classes of compounds. The drawback of on-column UV-detection in CE is sensitivity since only the capillary cross-section of 50–100 μm is available as path length. Instrumental attempts to increase the effective path length in capillaries include the Z-cell and the bubble cell [12]

The ability to measure at low wavelength without the interference of the absorption band of an organic modifier, the higher absorption of the analytes at wavelengths as low as 200 or 195 nm and the low dispersion and consequently higher peak make up for the reduced path length. Therefore, similar UV detection sensitivities can be obtained in CE versus HPLC (see Table 2.4). Nevertheless, concentration detection limits with UV detection are often insufficient for the direct analysis of most analytes in biological systems, making sample preconcentration steps such as stacking [13] – isotachophoresis [14] or derivatization prior to analysis necessary.

2.4.3.2 LIF detection

Since UV detection sensitivity was seen by many as insufficient for biological samples, considerable effort was put into developing and improving alternative detection capabilities (see Table 2.4). Detection principles, common to HPLC, were transferred to the special CE needs resulting from the characteristic capillary geometry. Fluorescence is among the most sensitive detection principles known. In a few reports, a deuterium or xenon arc lamp or a mercury line source was used for fluorescence excitation in CE [35, 36]. However, the small size of the detection window and the required high photon density for efficient excitation made lasers an ideal source for fluorescence detection in capillaries [37]. Laser induced fluorescence (LIF) detection became the preferred detection method in CE for high sensitivity applications. Figure 2.4 shows an schematic view of a commercial CE-LIF flow cell.

The quantum efficiency and molar extinction coefficient of the fluorophore, a good match between its absorption maximum and wavelength of the laser and the available laser power will determine the sensitivity of a LIF setup. Detection limits in the low nanomolar to picomolar range can be achieved [22, 38, 39]. Even single molecule detection is now feasible [40–42].

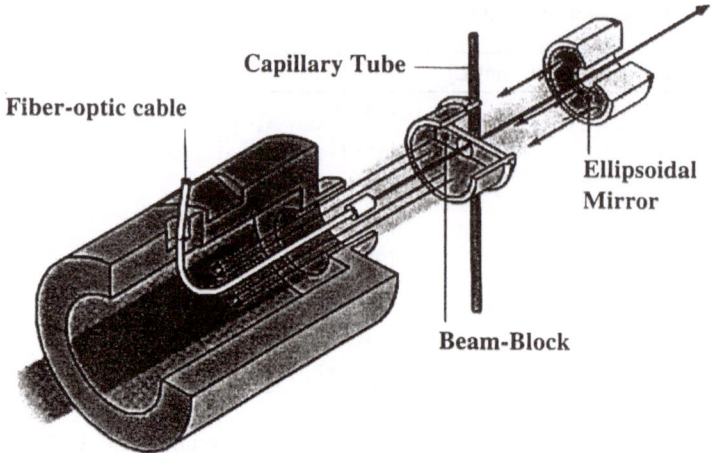

Figure 2.4 Schematic view of a CE-LIF detector (Beckman Instruments, Fullerton, CA). A fiber optic cable transmits laser light from the laser to the detector and illuminates a section of the capillary. Fluorescence is collected by the ellipsoidal mirror focused back onto the photomultiplier tube. To reduce unwanted laser light, a centered hole in the mirror allows most of the beam to pass. A beam block is used to attenuate scattered laser light. Reproduced with permission from Beckman Instruments.

Table 2.5 lists the most popular and economical laser light sources for CE-LIF applications. Ion lasers (Ar-ion, He-Cd, He-Ne) are relatively robust and economical lasers, available with an output power of 5–50 mW. A laser power of 1–5 mW at the point of detection is necessary for LIF in capillaries. It is also possible to use laser diodes with an emission wavelength in the red and infrared region of the spectrum [23]. Laser diodes exhibit an extreme longevity and robustness. The disadvantage of LIF is the requirement to tag the solutes of interest because most analytes do not show native fluorescence and/or can not be excited with

Table 2.5 Examples of laser light sources for detection with CE [43]

Laser source	wavelength	typical fluorophores
He-Cd (UV)	325 nm	coumarin dyes, OPA, ANTS
He-Cd (deep blue)	442 nm	NDA, acridine orange, green fluorescent protein mutants, CBQCA, NBD
Ar-Ion (blue)	488 nm	fluorescein-dyes
Ar-Ion (green	514 nm	rhodamine-dyes
He-Ne (green)	543 nm	phycoerythrin dyes, Ethidium bromide, Cy3
He-Ne (yellow)	594 nm	Texas Red dyes, YOYO-3,
He-Ne (orange)	612 nm	Cy5, TOTO-3
He-Ne (red)	633 nm	Cy5, allophycocyanin
laser diodes	635–1635 nm	Cy5, Cy7, indotricarbocyanine, La Jolla blue, IR dyes

commercial lasers. In addition, the optical properties of the label should closely match the excitation wavelength of the laser.

2.4.3.3 Indirect optical detection methods

Optical CE detection methods suffer not only from short optical path length but also from the fact, that the solutes of interest do not absorb the probing radiation. Although many solutes have some absorbance at low wavelength from 195–200 nm, small inorganic ions and organic solutes without double bonds, among them the majority of carbohydrates do not. In fluorescence detection, even fewer solutes exhibit natural fluorescence. To detect non absorbing and non fluorescent solutes with UV and LIF, respectively, indirect optical detection can be used [19–21, 24–26]. Figure 2.5 shows the principle: using a strong absorbing or fluorescent solute in the buffer, the detector is saturated, with the baseline at an elevated level. If a non-absorbing or non fluorescent solutes moves past the detector, it will displace molecules from the buffer thus generating a negative peak. In general, indirect detection methods are 1 to 2 order of magnitude less sensitive than direct methods. For best performance, the absorbing or fluorescent solute should act as the co-ion, should be ionized but with a low charge at the buffer pH, should have a high absorption coefficient and should have an electrophoretic mobility that is close to that of the solutes measured.

Indirect UV detection using sorbic acid has been used for the analysis of mono-, di- and trisaccharides [20]. Detection limits of 2×10^{-4} M have been reported. Although the detection sensitivity is not sufficient for the analysis of glycoproteins derived carbohydrates, indirect UV with sorbic acid works well for food analysis, for example the determination of glucose, fructose and saccharose in fruit juices [44].

Indirect LIF has been demonstrated with coumarin 343 as fluorophore for the analysis of monosaccharides [26]. However, the concentration detection limits were in the low micromolar range for saccharose, glucose and fructose and therefore only slightly better than with indirect UV detection.

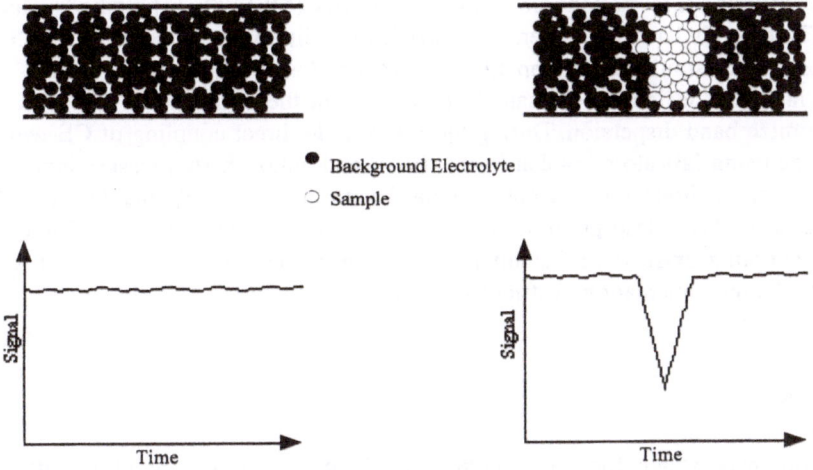

Figure 2.5 Principle of indirect detection methods

2.4.3.4 Electrochemical detection

Electrochemical detection for CE was first introduced by Wallingford and Ewing [29] using a porous glass joint to decouple the amperometric detector from the high electrical field of the separation capillary. With a high EOF, the solutes are pushed past the intersection towards the detector. Amperometric detection with a carbon electrode lends itself to the separation and sensitive detection of easily oxidized species such as catecholamines, achieving detection limits in the range of 10^{-8} M. Using different electrode materials such a copper [45–48] or gold [49, 50], the detection of carbohydrates is also feasible. Both constant electrode potential and pulsed operation have been described. Instrumental development includes the fabrication of microelectrodes and the use of normal size electrodes with a wall-jet approach. However, the low nanomolar detection limits could not be reached with carbohydrate solutes. Typically, a low micromolar sensitivity is reached. A severe disadvantage of electrochemical detection methods is at the time that no commercial instrumentation is on the market, inhibiting its widespread use.

2.4.3.5 Refractive index detection

Refractive index detection for CE based on interferometry has been demonstrated [51]. In one design using the off-axis method, the interference fringe pattern is generated with a laser beam roughly the size of the capillary. Because the round capillary acts as an optical element, light is reflected, refracted and transmitted when illuminating the detection window. In the far field, these rays form a fringe pattern which is altered by the refractive index of the solute inside the detection window. A second design uses a hologram to generate the interferogram. Both types of detection can be used with capillaries as small as 5–10 µm, but in terms of molar sensitivity, their performance is similar to indirect UV detection. As far electrochemical detection all reports with RI detection in CE rely on home built equipment.

2.4.3.6 Mass spectrometry detection

The advantages of coupling information rich mass spectrometry (MS) [33, 34] methods with highly efficient CE methods have long been recognized. In addition, the dimensions of CE and most MS methods are similar and compatible. In order to harvest the benefits of a CE-MS coupling, an interface design for the sample transfer from the liquid phase to the MS vacuum must minimize band dispersion. During the last decade direct coupling of CE with MS have been made using fast atom bombardment (FAB) and atmospheric pressure ionization (API) as ionization methods. Off-line coupling has been reported for plasma desorption (PD) and matrix assisted laser desorption ionization (MALDI). Although commercial interfaces for CE-MS coupling exist, the substantial capital investment and the high expertise, required to operate the instrumentation, restrict CE-MS to a small number of core labs.

2.5 CE modes

The term "electrophoresis" stands for a separation principle, based on differential migration of charged species in an electrical field. But similar to chromatography, it depends on the experimental conditions, which modulate the electrophoretic mobility and thus the separation. The electrophoretic mobility is a function of solute size and charge and in free solution, both have an effect on the separation. In biochemical analysis, solute mass is an important characteristic of a molecule and can be determined by electrophoretic separation in a gel network. In isoelectric focusing, the influence of the solutes charge can be made dominant over size allowing for isoelectric point measurements. In some CE techniques feasible on commercial CE instrumentation also chromatographic interactions are implemented. Today, CE consists of a family of 6 different modes, which are summarized in Table 2.6. Their mode of operation is depicted in Figure 2.6.

Table 2.6 Modes of Capillary Electrophoresis

Mode	Abbreviation	Separation Principle
Capillary Electro Chromatography	CEC	chromatographic interaction
Micellar Electrokinetic Capillary Chromatography	MECC	partition between micellar and buffer phase
Capillary Zone Electrophoresis	CZE	electrophoretic mobility
Isotachophoresis	ITP	electrophoretic mobility
Capillary Isoelectric Focusing	CIEF	isoelectric point
Capillary Gel Electrophoresis	CGE	size

2.5.1 Capillary Electro Chromatography

Capillary Electro Chromatography (CEC) [52–55], pioneered by the group of John Knox, is basically a chromatographic separation technique. A stationary phase, typically silica or chemically modified silica, is used as chromatographic support, with particle diameters of 1–3 µm and filled into capillaries of 50–200 µm inner diameter. The mobile phase, which has to contain some electrically conductive component, is moved by the EOF when an electrical field is applied along the stationary phase. As in chromatography, the separation mechanism depends on the nature of the stationary phase, e.g. hydrophobic interaction with reversed phase materials or electrostatic interaction with ion exchange phases. Because of the plug-shaped flow profile of the EOF, plate heights are smaller and the separation power of CEC is superior to comparable LC systems.

2.5.2 Micellar Electrokinetic Capillary Chromatography

Micellar Electrokinetic Capillary Chromatography (MECC) conceived by Shigeru Terabe is also a hybrid of chromatography and electrophoresis [16, 56]. This technique grew out of the dilemma that electrically neutral substances could not be separated by electrophoresis. But by adding a micelle-forming detergent to the buffer such as sodium dodecyl sulfate (SDS) in a concentration above the critical micellar concentration, the micellar phase acts very similar to a stationary phase in chromatography. The interior of the micelle consists of hydrophobic

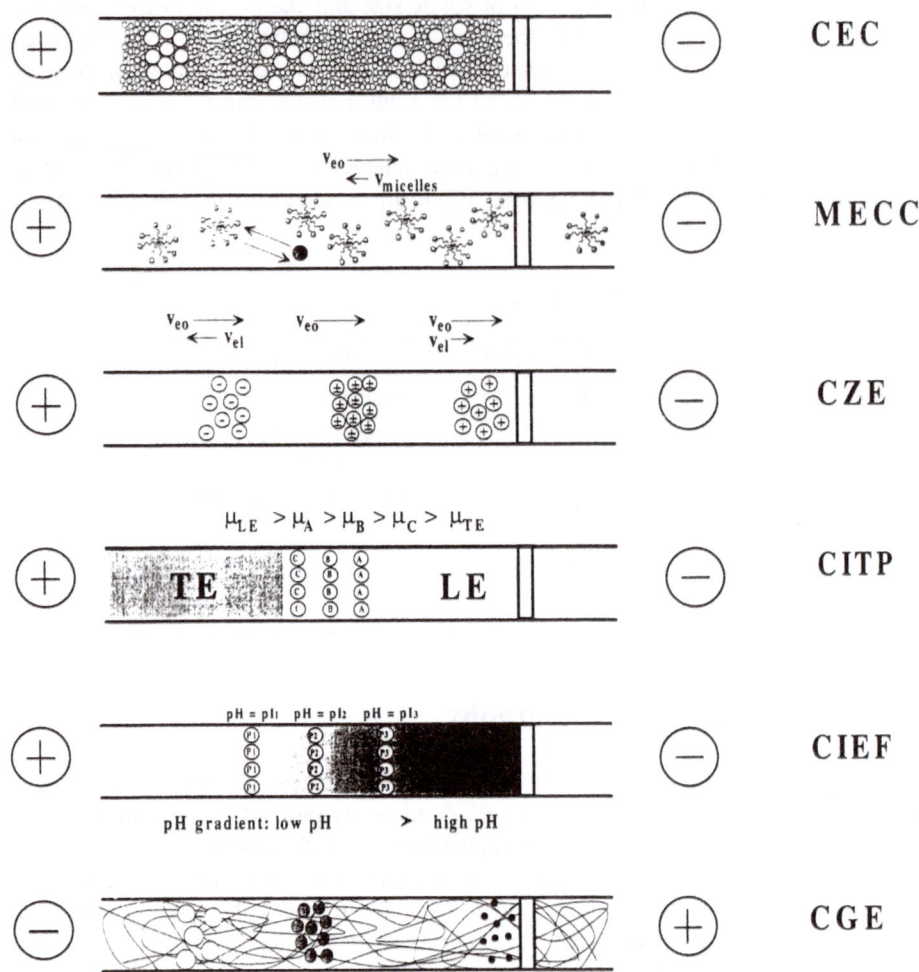

Figure 2.6 CE Modes

alkyl chains and the surface of negative hydrophilic sulfate groups. The highly negatively charged micelles of SDS migrate against the EOF, but since their electrophoretic velocity is smaller than the EOF, they still are swept to the detector end of the capillary. The situation is reversed for positively charged micelles.

The separation is based on differential residence times of the solutes in the micellar and the buffer phase. Hydrophilic solutes will reach the detector with the EOF, strongly hydrophobic solutes will spend more time inside the micellar phase and are therefore retarded. The separation window is defined by the migration time of a neutral marker for the EOF and the migration time of the micelles. MECC separations are very similar to reversed phase chromatography. Particularly for solutes with molecular weights below 1000, MECC provides separations as good as HPLC or better, yet with very simple and rapid method development.

MECC was successfully applied to the separation of a wide variety of compounds, including amino acids, peptides, aflatoxins, organic halogen compounds of technical origin as well as various drugs suchs as vitamines, barbiturates, steroids, antibiotics and anti-inflammatory drugs. In addition, MECC with chiral additives proved useful in enatiomeric separations. Other micelle forming compounds such as sodium cholate or cetyltrimethyl ammonium bromide (CTAB) have also been used.

2.5.3 Capillary Zone Electrophoresis

Capillary Zone Electrophoresis (CZE) [3, 35] is CE performed in a buffer filled capillary and was developed by Jim Jorgenson and his group in the early 1980ties. The separation power of several 100.000 plates in a miniaturized automated system was recognized by other researchers in the field and eventually instrument manufacturers and paved the way for the rapid growth of CE during the last 8–10 years. CZE is together with MECC the most popular and most widely used CE technique. CZE uses differences in charge and size of the solutes for the separation in an electrical field. As with MECC and CEC, the EOF serves as a pump, directing a constant flow to the detector. Both positive and negative solutes are separated with CZE, the positive ones eluting in front of the EOF, the negative solutes being retarded compared to the EOF. CZE demonstrated excellent results with peptides, tryptic maps and small oligonucleotides [57]. Proteins are also well separated if any chromatographic interaction with the fused-silica surface can be avoided.

2.5.4 Isotachophoresis

Isotachophoresis (ITP) is the oldest of the CE methods, presented here. Its instrumental development was particularly advanced by Frans Everaerts and his group [14]. ITP is comparable to chromatographic focusing: the solutes to be separated form discrete zones according to their electrophoretic velocity, between a leading (LE) and a terminating (TE) electrolyte. This train of solute bands moves at constant velocity pass the stationary detector. Every solute has a specific electrophoretic mobility, but field strength changes along the solute train. If zone mixing due to diffusion occurs at the front or the end of a band, these solutes leaving "their" zone will experience a different electrical field, which causes a slow down or an acceleration. The resulting self-focusing, the most important advantage of ITP, is most often applied to ion analysis and charged small organic molecules, but failed to gain widespread acceptance. However, isotachophoretic effects may be useful for sample concentration during CZE injections.

2.5.5 Capillary Isoelectric Focusing

Isoelectric focusing (IEF) [58] in capillaries is a two-step process. In the first step, the focusing, the sample is mixed with the so-called ampholines (polymeric zwitterions having different pK_a's) and filled into the capillary. Upon application of an electrical field, a pH gradient is formed through the ampholines. The sample solutes migrate in this gradient to the point at which they are electrically neutral and therefore experience no acceleration from the field (isoelectric point, pI) In the second step, the stationary state is mobilized in order to transport the solutes past the detector. This can be done with a micro pump or, more elegantly, by exchanging the cathode buffer and inducing a pH jump. CIEF is particularly suited to the separation of proteins, sometimes on a micro preparative scale. A prerequisite for the formation of the stationary state and thus a requirement for CIEF, is the complete lack of EOF in the capillary during the focusing step.

2.5.6 Capillary Gel Electrophoresis

The last CE separation mode in this series is Capillary Gel Electrophoresis (CGE) [59]. In this mode, cross-linked gels or entangled polymer solutions in capillaries act as anti-convective media to completely suppress the EOF and to keep band diffusion small. In addition, sieving occurs as the solutes move through the pore network of the gels or entangled polymer solution. By choosing appropriate separation conditions (buffer-pH, buffer additives, polymer composition) the electrophoretic mobility can be made to depend only on molecular size. Therefore, CGE can be used for size separations. CGE is most applicable in biochemical analysis for the separation of oligonucleotides, DNA digests and SDS protein complexes. CGE represents a miniaturized automated version of slab gel electrophoresis and will potentially have a major impact on routine analytical procedures in biotechnology.

References

1. M. Novotny, "Liquid chromatography in columns of capillary dimensions", in "Microcolumn high performance liquid chromatography". (ed. P. Kucera), Elsevier, **1984**, Amsterdam,194-259.

2. V. Pretorius, B. Hopkins, and J.D. Schieke, "A new concept for high speed liquid chromatography", *J. Chromatogr.*, *99* **1974** 23-30.

3. J.W. Jorgenson, and K.D. Lukacs, "Zone electrophoresis in open tubular glass capillaries", *Anal. Chem.*, *53* **1981** 1298-1302.

4. S. Hjerten, "Free zone electrophoresis", *Chromatogr. Rev.*, *9* **1969** 122-219.

5. R.J. Nelson, A. Paulus, A.S. Cohen, A. Guttman, and B.L. Karger, "Use of Peltier thermoelectric devices to control temperature in high performance capillary electrophoresis", *J. Chromatogr.*, *480* **1989** 111-127.

6. R.J. Wieme, "Theory of electrophoresis", in "Chromatrography: A Lab Handbook of Chromatography and Electrophoresis Methods", (ed. E. Heftmann), Van Norstrand, **1984**, New York, 228-278.

7. A. Klockow, R. Amado, H.M. Widmer, and A. Paulus, "The influence of buffer composition on separation efficiency and resolution in capillary electrophoresis of 8-aminonaphthalene-1,3,6-trisulfonic acid labeled monosaccharides and complex carbohydrates", *Electrophoresis*, *17* **1996** 110-119.

8. J.C. Giddings, "Unified Separation Science", John Wiley & Sons, Inc., **1991**, New York.

9. R.G. Brownlee, and S.W. Compton, "Automated instrumentation for analytical capillary electrophoresis", *Am. Lab.*, *20* **1988** 10-17.

10. R.J. Nelson, and D.S. Burgi, "Temperature control in capillary electrophoresis", in "Handbook of Capillary Electrophoresis", (ed. J.P. Landers), CRC Press, **1994**, Boca Raton, 549-562.

11. A. Cifuentes, W.T. Kok, and H. Poppe, "Capillary electrophoresis using air and helium as cooling fluids", *J. Microcolumn Separations*, *7* **1995** 365-374.

12. I.H. Grant, and W. Steuer, "Extended pathlength UV absorbance detector for capillary zone electrophoresis", *J. Microcolumn Sep.*, *2* **1990** 74.

13. R.L. Chien, and D.S. Burgi, "Sample stacking of an extremly large injection volume in high performance capillary electrophoresis", *Anal. Chem.*, *64* **1992** 1046.

14. E.E.P. Mikkers, F.M. Everarts, and T.E.P.M. Verheggen, "Isotachophoresis: Theory, Instrumentation and Application", *J. Chromatogr.*, *169* **1979** 11-20.

15. A.E. Bruno, E. Gassmann, N. Perikles, and K. Anton, "On-column capillary flow cell utilizing optical waveguides for chromatographic applications", *Anal. Chem.*, *61* **1989** 876.

16. S. Terabe, K. Otsuka, K. Ichikawa, A. Tsuchiya, and T. Ando, "Electrokinetic separations with micellar solution and open-tubular capillaries", *Anal. Chem.*, *56* **1984** 111.

17. F. Foret, M. Deml, V. Kahle, and P. Bocek, "On-line fiber optic UV detection cell and conductivity cell for capillary zone electrophoresis", *Electrophoresis*, *7* **1986** 430.

18. G.J.M. Bruin, G. Stegeman, A.C. van Asten, X. Xu, J.C. Kraak, and H. Poppe, "Optimization and evaluation of the performance of arrangements for UV detection in high resolution separations using fused silica capillaries", *J. Chromatogr.*, *559* **1991** 163.

19. F. Foret, S. Fanali, L. Ossicini, and P. Bocek, "Indirect photometric detection in capillary zone electrophoresis", *J. Chromatogr.*, *470* **1986** 299.

20. P.J. Oefner, A.E. Vorndran, E. Grill, C. Huber, and G.K. Bonn, "Capillary zone electrophoretic analysis for carbohydrates by direct and indirect UV detection", *Chromatographia*, *34* **1992** 308-316.

21. G.J.M. Bruin, A.C. van Asten, X. Xu, and H. Poppe, "Theoretical and experimental aspects of indirect detection in capillary electrophoresis", *J. Chromatogr.*, *608* **1992** 97.

22. Y. Chen, and N.J. Dovichi, "Subattomole amino acid analysis by capillary zone electrophoresis and laser-induced fluorescence", *Science*, *242* **1988** 562-564.

23. F.-T.A. Chen, A. Tusak, S.L. Pentoney Jr., K. Konrad, C. Lew, E. Koh, and J. Sternberg, "Semiconductor laser-induced fluorescence detection in capillary electrophoresis using a cyanine dye", *J. Chromatogr. A*, *652* **1993** 355-360.

24. W. Kuhr, and E.S. Yeung, "Optimization of sensitivity and separation in capillary zone electrophoresis with indirect fluorescence detection", *Anal. Chem.*, *60* **1988** 2642.

25. L. Gross, and E.S. Yeung, "Indirect fluorometric detection of cations in capillary zone electrophoresis", *Anal. Chem.*, *62* **1990** 427.

26. T.W. Garner, and E.S. Yeung, "Indirect fluorescence detection of sugars separated by capillary zone electrophoresis with visible laser excitation", *J. Chromatogr.*, *515* **1990** 639.

27. X. Huang, T.K. Pang, M.J. Gordon, and R.N. Zare, "On-column conductivity detector for capillary zone electrophoresis", *Anal. Chem.*, *59* **1987** 2747-2749.

28. X. Huang, J.A. Luckey, M.J. Gordon, and R.N. Zare, "Quantitative analysis of low molecular weight carboxylic acids by capillary zone electrophoresis", *Anal. Chem.*, *61* **1989** 766-770.

29. R.A. Wallingford, and A.G. Ewing, "Capillary zone electrophoresis with electrochemical detection", *Anal. Chem.*, *59* **1987** 1762-1766.

30. T.M. Olefirowicz, and A.G. Ewing, "Capillary Electrophoresis in 2 and 5 micron diameter capillaries: application to cytoplasma analysis" *Anal.Chem.*, *62* **1990** 1872-1876.

31. W. Yu, and N.J. Dovichi, "Attomole amino acid determination by capillary zone electrophoresis with thermooptical absorbance detection", *Anal. Chem.*, *61* **1989** 37.

32. A.E. Bruno, A. Paulus, and D.J. Bornhop, "Thermooptical absorbtion detection in 25-mm-i.d. capillaries: capillary electrophoresis of dansyl amino acid mixtures", *Appl. Spectrosc.*, *45* **1991** 462-467.

33. J. Cai, and J. Henion, "Capillary electrophoresis-mass spectrometry", *J. Chromatogr.* A, *703* **1995** 667-692.

34. R.D. Smith, D.R. Goodlett, and J.H. Wahl, "Capillary electrophoresis – mass spectrometry", in "Handbook of capillary electrophoresis", (ed. J.P. Landers), CRC Press, **1994**, Boca Raton,185-206.

35. J.W. Jorgenson, and K.D. Lukacs, "Capillary zone electrophoresis", *Science*, *222* **1983** 266-272.

36. M. Albin, R. Weinberger, E. Sapp, and S. Moring, "Fluorescence detection in capillary electrophoresis: evaluation of derivatizing reagents and techniques", *Anal. Chem.*, *63* **1991** 417-421.

37. E. Gassman, J.E. Kuo, and R.N. Zare, "Electrokinetic separation of chiral compounds", *Science*, *230* **1985** 813-814.

38. H. Swerdlow, J.Z. Zhang, D.Y. Chen, H.R. Harke, R. Grey, S. Wu, C. Fuller, and N.J. Dovichi, "Three DNA sequencing methods using capillary gel electrophoresis and laser-induced fluorescence", *Anal. Chem*, *63* **1991** 2835-2841.

39. J.V. Sweedler, J.B. Shear, H.A. Fishman, R.N. Zare, and R.H. Scheller, "Fluorescence detection in capillary zone electrophoresis using a charged coupled device with time-delayed integration", *Anal. Chem.*, *63* **1991** 496-502.

40. D.T. Chiu, A. Hsiao, A. Gaggar, R.A. Garza-Lopez, O. Orwar, and R.N. Zare, "Injection of ultrasmall samples and single molecules into tapered capillaries", *Anal. Chem.*, *69* **1997** 1801-1807.

41. W. Tan, and E.S. Yeung, "Monitoring the reactions of single enzymes molecules and single metal ions", *Anal. Chem.*, *69* **1997** 4242-4248.

42. A. Castro, and J.G.K. Williams, "Single-molecule detection of specific nucleic acid sequences in unamplified genomic DNA", *Anal. Chem.*, *69* **1997** 3915-3920.

43. R.P. Haugland, "Handbook of fluorescent probes and research chemicals", Molecular Probes, **1996**, Eugene.

44. A. Klockow, A. Paulus, V. Figueiredo, R. Amado, and H.M. Widmer, "Determination of carbohydrates in fruit juices by liquid chromatography", *J. Chromatogr.*, A, *680* **1994** 187-200.

45. L.A. Colon, R. Dadoo, and R.N. Zare, "Determination of carbohydrates by capillary zone electrophoresis with amperometric detection at a copper microelectrode", *Anal. Chem.*, *65* **1993** 476-481.

46. J. Ye, and R.P. Baldwin, "Determination of carbohydrates , sugar acids and alditols by capillary electrophoresis and electrochemical detection at a copper electrode", *J. Chromatogr.* A, *687* **1994** 141-148.

47. X. Huang, and W.T. Kok, "Determination of sugars by capillary electrophoresis using cuprous oxide modified electrodes", *J. Chromatogr.* A, *707* **1995** 335-342.

48. J. Ye, and R.P. Baldwin, "Amperometric detection in capillary electrophoresis with normal size electrodes", *Anal. Chem.*, *65* **1993** 3525-3527.

49. R.M. Cassidy, W. Lu, and V.-P. Tse, "Auxiliary Electroosmotic Flow for Postcapillary Reaction Detection in Capillary Electrophoresis", *Anal. Chem.*, *66* **1994** 2578-2583.

50. T.J. O'Shea, S.M. Lunte, and W.R. LaCourse, "Detection of carbohydrates by capillary electrophoresis with pulsed amperometric detection", *Anal. Chem.*, *65* **1993** 948-951.

51. A.E. Bruno, B. Krattiger, F. Maystre, and H.M. Widmer, "On-column laser-based refractive index detector for capillary electrophoresis", *Anal. Chem.*, *63* **1991** 2689-2697.

52. J.H. Knox, and I.H. Grant, "Miniaturization in pressure and electroendoosmotically driven liquid chromatography: some theoretical considerations", *Chromatographia*, *24* **1987** 135-143.

53. B. Behnke, E. Grom, and E. Bayer, "Evaluation of the parameters determining the performance of electrochromatography in packed capillary columns", *J. Chromatogr. A*, *716* **1995** 207-213.

54. M.T. Dulay, C. Yan, D.J. Rakesraw, and R.N. Zare, "Automated capillary electrochromatography: reliability and reproducibility studies", *J. Chromatogr. A*, *725* **1996** 361-366.

55. N.W. Smith, and M.B. Evans, "The efficient analysis of neutral and high polar pharmaceutical compounds using reversed-phase and ion-exchange electrochromatography", *Chromatographia*, *41* **1995** 197-203.

56. M.G. Khaledi, "Micellar electrokinetic capillary chromatography", in "Handbook of capillary electrophoresis", (ed. J.P. Landers), CRC Press, **1994**, Boca Raton,43-93.

57. J.P. Landers, "Handbook of capillary electrophoresis", CRC Press, **1994**, Boca Raton.

58. F. Kilar, "Isoelectric focusing in capillaries", in "Handbook of capillary electrophoresis", (ed. J.P. Landers), CRC Press, **1994**, Boca Raton,95-109.

59. C. Heller, "Analysis of Nucleic Acids by Capillary Electrophoresis", Vieweg Publishing, **1997**, Wiesbaden.

3 Structures and properties of carbohydrates

The expression „carbohydrate" reflects the belief in the 19th century, that this class of compounds with the general formula $C_n(H_2O)_n$ are hydrates of carbon. This early definition has been widened to any multifunctional compounds containing a number of hydroxyl groups of similar reactivity and at least one asymmetric carbon atom. Consequently, a number of examples exist that have very different molecular formulas but are nevertheless due to their chemical reactivity carbohydrates, for example deoxysugars or aminosugars.

Amino acids form polymer structures by forming peptidic bonds, a head-to-tail condensation of the amino and the carboxy group. Two amino acids can form just two different dipeptides. Monosaccharides can be joined together via all equivalent hydroxy groups. The complicated stereochemistry adds to the variability. Therefore two identical monosaccharides can form up to 11 different disaccharides. Whereas 4 different amino acids can form 4 tetrapeptides, 4 different hexoses (e.g. glucose, galactose, mannose, gulose) can produce more than 36.000 tetrasaccharides [1]. The consequence is an enormous diversity and complexity in carbohydrate structure and chemistry.

Carbohydrates can be divided into monosaccharides, oligosaccharides and polysaccharides depending on the number of sugar units that are linked together. Whereas monosaccharides can not be hydrolyzed to smaller molecules the oligosaccharides contain 2–10 monosaccharide residues joined by glycosidic linkages. The polysaccharides are polymers of high molecular weight consisting of more than 10 monosaccharide units with the majority of the naturally occurring polysaccharides containing 80–100 units. Due to their size polysaccharides are less water soluble compared to the mono- and oligosaccharides, they do not taste sweet and chemically they are less reactive.

3.1 Monosaccharides

In principal the monosaccharides are polyhydroxyaldehydes (aldoses) or polyhydroxyketones (ketoses) deriving from glyceraldehyde or dihydroxyacetone respectively through insertion of CHOH-units. According to the number of carbon atoms the aldoses are classified, starting with the triose glyceraldehyde, in tetroses, pentoses (e.g. ribose, xylose) and hexoses (e.g. glucose, mannose, galactose). The ketoses, starting with the triulose dihydroxyacetone, are called tetruloses, pentuloses (e.g. ribulose, xylulose) and hexuloses (e.g. fructose, sorbose).

Glyceraldehyde has one asymmetric carbon atom resulting in an enantiomeric pair, the D- and the L-glyceraldehyde. The prefixes D and L designate the absolute configuration. The compounds deriving from D-glyceraldehyde are called D-aldoses and those deriving from L-glyceraldehyde are called L-aldoses. For monosaccharides with more than one asymmetric carbon atom, the symbols D and L refer to the absolute configuration of the asymmetric carbon farthest away from the aldehyde group. A molecule with n asymmetric centers has 2^n stereoisomers. Therefore insertion of CHOH-groups in D- and L-glyceraldehyde ($n = 1$) leads to formation of four aldotetroses ($n = 2$), eight pentoses ($n = 3$) and sixteen isomeric hexoses ($n = 4$). Monosaccharides which differ only in the configuration at one single carbon atom are called *epimers*. D-Glucose and D-mannose for example are 2-epimers whereas D-glucose and D-galactose are 4-epimers. Like for the aldoses the same series of D- and L-ketoses is derived through insertion of CHOH-groups starting with the enantiomeric D- and L-tetruloses. For a systematic representation of the stereochemical relations of aldoses and ketoses („sugar-trees") the authors refer to references [2–4].

Starting with tetrose and 2-pentulose, all monosaccharides can cyclize into rings forming an intramolecular hemiacetal or hemiketal. The resulting five and six membered lactols are called *furanoses* and *pyranoses*, respectively. The lactol formation leads to an additional chiral center and therefore formation of two diastereomeric pyranoses and furanoses each. These diastereomers are called *anomers* and are designated with α and β. α means that the hydroxyl group attached to C1 of the pyranose is below the plane of the ring; β means that it is above the plane of the ring. The same nomenclature applies to the furanose form e.g. of fructose, except that α and β refers to the hydroxyl group attached to C2 which is the anomeric carbon atom in furanoses.

Consequently in solution each monosaccharide can occur in five different forms, one open-chain and four ring forms (Figure 3.1). Due to the great tendency of the monosaccharides to cyclize only little of the open-chain form is present. In equilibrium glucose is present almost exclusively in its pyranose forms with 36% α and 64% β [4]. For other monosaccharides this equilibrium mixture can be very different. The different ring forms always interconvert through the open chain form of the sugar. This change is called mutarotation since it goes along with a change in the optical rotation of the solution of a distinct monosaccharide. In the following the term for a single monosaccharide such as glucose or fructose always refers to its equilibrium mixture of the open-chain and the ring forms. For a comprehensive discussion of the structural and stereochemical features of monosaccharides the authors refer to general biochemistry and organic chemistry textbooks [2–6].

Monosaccharides containing functional groups other than those found in neutral monosaccharides such as acidic sugars, aminosugars and deoxysugars are important components in oligo- and polymeric carbohydrates.

ACIDIC SUGARS: There are three different types of acidic sugars which are classified according to which terminal group has been oxidized. *Aldonic acids* are formed by oxidation of the carbonyl group, *uronic acids* by oxidation of the terminal alcohol group and *aldaric acids* by oxidation of both the carbonyl and the terminal alcohol group. The most important type of acidic sugars are the uronic acids, which occur almost exclusively as hexuronic acids. They are intermediates in the conversion of hexoses and pentoses in biosynthetic pathways. D-Glucuronic acid and D-galacturonic acid are found in plant gums and bacterial cell walls whilst D-glucuronic acid and L-iduronic acid are components of proteoglycans.

AMINOSUGARS: The anomeric carbon atom of a monosaccharide can be linked to the nitrogen of an amine by an N-glycosidic bond. Most aminosugars occur as constituents in oligo- and polysaccharides. The most abundant aminosugar is 2-amino-2-deoxy-D-glucose which, as its N-acetylated derivative (N-acetylglucosamine), occurs widely in polysaccharides, glycoproteins and proteoglycans. 2-Acetamido-2-deoxy-D-galactose (N-acetylgalacto-

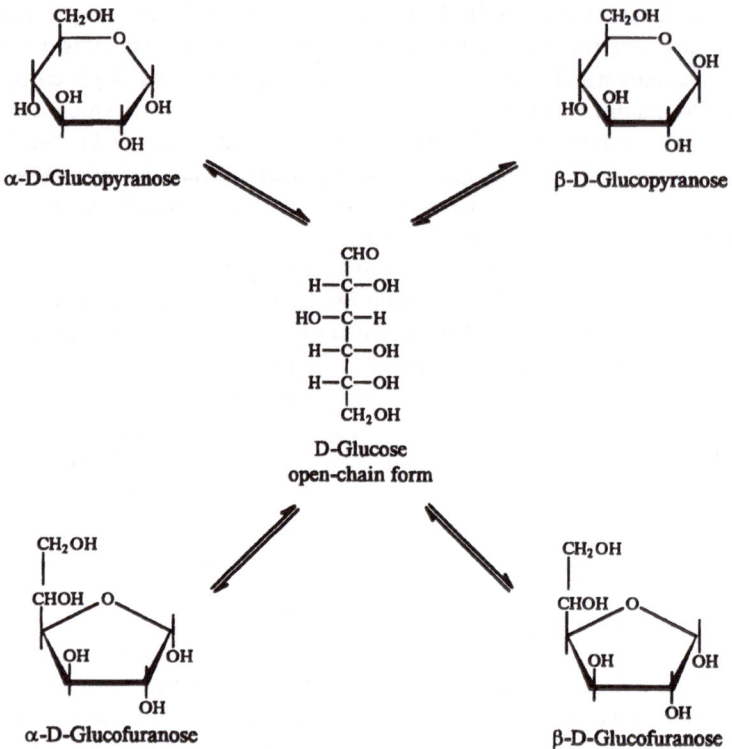

Figure 3.1 Structures of the five different species of D-glucose in water

samine) and -D-mannose as well as the two higher amino sugars 5-acetamido-3,5-dideoxy-D-glycero-D-galacto-2-nonulopyranonic acid (N-acetylneuraminic acid) and 5-glycolylami-do-3,5-dideoxy-D-glycero-D-galacto-2-nonulopyranonic acid (N-glycolylneuraminic acid) are also found in glycoproteins.

DEOXYSUGARS: Deoxysugars, in which one or more of the hydroxyl groups of the aldose or ketose has been replaced by hydrogen, are widely distributed in nature. 6-Deoxyhexoses such as 6-deoxy-L-mannose (L-rhamnose) and 6-deoxy-L-galactose (L-fucose) are present in many polysaccharides, glycoproteins and plant glycosides. 2-deoxy-D-ribose, linked over two diesterbonds to phosphoric acid, forms the backbone of DNA (deoxy ribonucleic acid).

3.2 Oligosaccharides

The linkage of two monosaccharides through an O-glycosidic bond results in the formation of a disaccharide. Glycosidic linkage between the lactol groups (OH-group at the anomeric carbon) of two monosaccharides results in a non reducing disaccharide such as saccharose, whilst the participation of one lactol and one free hydroxyl group to build the linkage results in a reducing disaccharide (e.g. maltose, lactose). Through elongation of the disaccharide 'chain' via O-glycosidic bonds oligosaccharides and polysaccharides are formed. The manner in which the various monosaccharides are joined is very complex due to the great number of possible substitution positions. Additionally, branched structures do exist because of disubstitution of a single monosaccharide residue. Therefore to fully define the primary structure of oligo- and polysaccharides, the identity of all monosaccharide residues, the sequence of these residues, their position and anomeric configuration, and the position of any other substituent has to be determined [7]. The secondary structure or conformation of oligo- and polysaccharides is given by the relative orientation of the component residues (i. e. rotation) about the glycosidic linkage. This relative orientation is described by the torsion angles ϕ, ψ and ω of the glycosidic bond between two carbohydrate residues as depicted in Figure 3.2 [4, 7].

To describe the composition of oligosaccharides a reduced nomenclature is applied which uses a short name consisting of three letters for the monosaccharide and the suffixes f and p for furanose and pyranose. Table 3.1 shows a list of the monosaccharide short names which are used throughout this book. If no other configuration is mentioned, the D-configuration of the sugar is assumed. A selection of common di- and oligosaccharides plus their structure and source is given in Table 3.2.

The most common naturally occurring oligosaccharide is saccharose which is widely distributed throughout the plant world. It is the main soluble carbohydrate reserve and an important energy source in human diet. The principle sources for the commercial production of saccharose are sugar cane, sugar beet and the sap of maple trees. Lactose is mainly found in mammalian milk with 2.0–8.5% depending on the mammal [7]. It is present either as the free disaccharide or combined with other residues such as GlcNAc, L-Fuc, Gal and NeuAc. The concentration of these higher oligosaccharides is between 0.3 and 0.6%. Industrially lactose is prepared from whey. A special family of oligosaccharides is the raffinose family. Raffinose is a naturally occurring trisaccharide which is as widely distributed as saccharose but in much lower concentrations (e.g. 0.05% in sugar beet, 1.9% in soybean). From its structure it

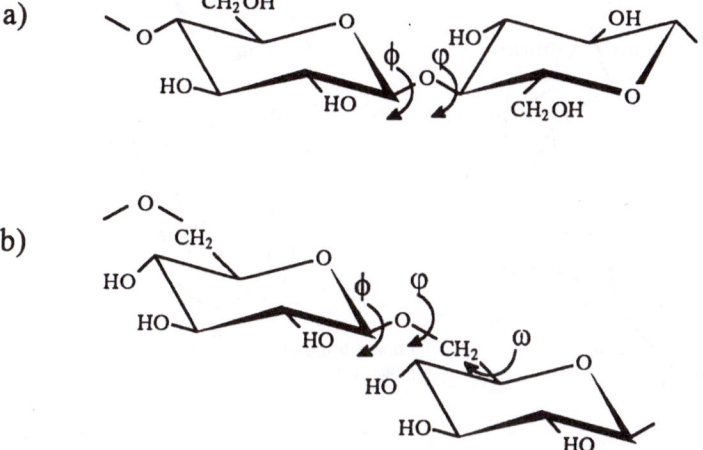

Figure 3.2 Rotation of individual residues about the glycosidic linkage in a) cellulose and b) dextran

Table 3.1 Monosaccharide short names

Monosaccharide	Short name	Monosaccharide	Short name
Glucose	Glc	Glucuronic acid	GlcA
Galactose	Gal	Galacturonic acid	GalA
Mannose	Man	N-acetylglucosamine	GlcNAc
Fructose	Fru	N-acetylgalactosamine	GalNAc
Arabinose	Ara	N-acetylneuraminic acid	NeuAc
Xylose	Xyl	N-glycolylneuraminic acid	NeuGc
Fucose	Fuc	Iduronic acid	IdoA
Idose	Ido		
Rhamnose	Rha		

can be regarded as an O-galactosylsaccharose (see Table 2.3) which is followed by stachyose with one and verbascose with two additional galactose residues. These oligosaccharides are synthesized from saccharose.

The highest molecular weight oligosaccharides which have been obtained in crystalline form are the cyclomaltoses or cyclodextrins. These cyclomalto-oligosaccharides have been prepared by enzymatic digestion of starch by a *Bacillus macerans* enzyme and include cyclomaltohexaose, cyclomaltoheptaose and cyclomaltooctaose. Due to their cyclic structure the cyclodextrins can bind small molecules to form inclusion complexes (Figure 3.3). Since the included molecules are protected in this way cyclodextrins are used in the production of pharmaceuticals, pesticides, foodstuffs and toiletries as protecting agents and also as emulsion stabilizers.

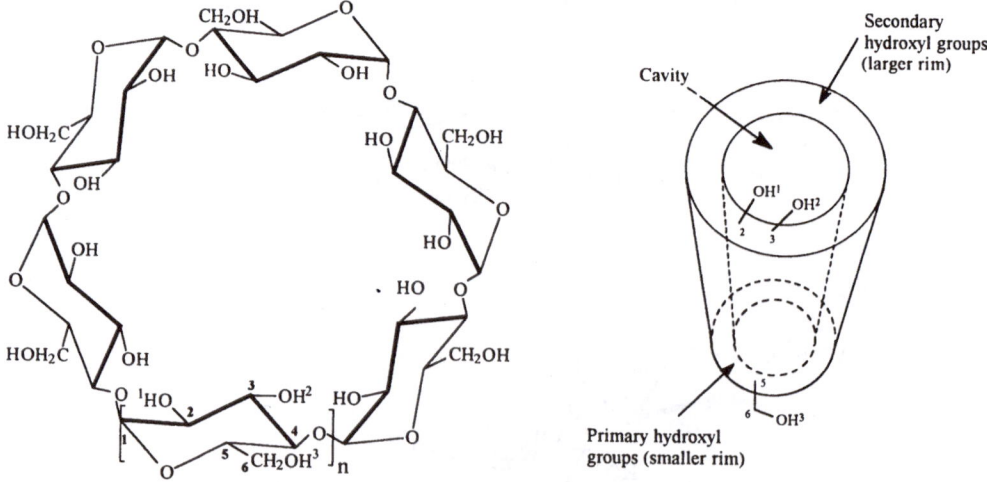

Figure 3.3 Structure of cyclodextrins; $n = 0$, 1 or 2 for α-, β- and γ-cyclodextrin

Table 3.2 Structure and source of common oligosaccharides

Trivial name	Structure	Source
Disaccharides		
Cellobiose	O-β-D-Glcp-(1,4)-D-Glcp	Cellulose
Chitobiose	O-β-D-GlcpNAc-(1,4)-D-GlcpNAc	Chitin
Isomaltose	O-α-D-Glcp-(1,6)-D-Glcp	production of glucose from starch
Lactose	O-β-D-Galp-(1,4)-D-Glcp	milk
Maltose	O-α-D-Glcp-(1,4)-D-Glcp	starch, sugar beet, honey
Melibiose	O-α-D-Galp-(1,6)-D-Glcp	cacao
Neohesperidose	O-α-L-Rhap-(1,2)-D-Glcp	glycosides
Nigerose	O-α-D-Glcp-(1,3)-D-Glcp	honey, beer
Rutinose	O-α-L-Rhap-(1,6)-D-Glcp	glycosides
Saccharose	O-β-D-Fruf-(2,1)-α-D-Glcp	sugar beet, sugar cane, maple tree
Trehalose	O-α-D-Glcp-(1,1)-α-D-Glcp	mushrooms
Trisaccharides		
Fucosidolactose	O-α-D-Fucp-(1,2)-O-β-D-Galp-(1,4)-D-Galp	human milk
Kestose	O-α-D-Glcp-(1,2)-O-β-D-Fruf-(6,2)-β-D-Fruf	honey
Maltotriose	O-α-D-Glcp-(1,4)-O-α-D-Glcp-(1,4)-D-Glcp	degradation of starch
Manninotriose	O-α-D-Galp-(1,6)-O-α-D-Galp-(1,6)-D-Glcp	manna
Melezitose	O-α-D-Glcp-(1,3)-O-β-D-Fruf-(2,1)-α-D-Glcp	manna, nectar
Raffinose	O-α-D-Galp-(1,6)-O-α-D-Glcp-(1,2)-β-D-Fruf	sugar beet, sugar cane
Tetrasaccharides		
Maltotetraose	O-α-D-Glcp-(1,4)-O-α-D-Glcp-(1,4)-O-α-D-Glcp-(1,4)-D-Glcp	degradation of starch
Stachyose	O-α-D-Galp-(1,6)-O-α-D-Galp-(1,6)-O-α-D-Glcp-(1,2)-β-D-Fruf	legume seeds, artichokes
Higher oligosaccharides		
Verbascose	O-α-D-Galp-(1,6)-O-α-D-Galp-(1,6)-O-α-D-Galp-(1,6)-O-α-D-Glcp-(1,2)-β-D-Fruf	legume seeds
Maltopentaose	[O-α-D-Glcp-(1,4)]$_4$- D-Glcp	degradation of starch
Cyclodextrins	Cyclo-[O- α-D-Glcp-(1,4)]$_6$	action of *Bacillus macerans* on starch
	Cyclo-[O- α-D-Glcp-(1,4)]$_7$	
	Cyclo-[O- α-D-Glcp-(1,4)]$_8$	

3.3 Polysaccharides

Polysaccharides consist of monosaccharides linked through O-glycosidic bonds. Polysaccharides are built from either one (homopolysaccharides) or several monosaccharide types (heteropolysaccharides) which can be linked linear (cellulose, amylose) or branched to different extents (amylopectin, glycogen, guaran). Additionally, the sequence of the monosaccharide residues can be

(1) periodic, with one period comprising one or more residues (cellulose, amylose),

(2) periodic over certain sections of different length which are separated through sections with aperiodic sequences (algine, carrageenan, pectine),

(3) aperiodic throughout (carbohydrate part of glycoproteins).

Periodic sequences in the primary structures of polysaccharides lead to regular patterns in secondary structures which lead to sterically regular gross conformations. In contrast irregularities in primary and secondary structures as well as large branched structures inhibit formation of tertiary structures. This is important since the structure of polysaccharides affects their solubility. For example cellulose and amylose are insoluble in their natural state due to their highly ordered conformation, the resulting ordered packing and the interactions between the polysaccharide chains. Therefore aggregation of chains and precipitation are favored. On the other hand branched polysaccharides such as amylopectin or glycogen are more soluble since less interaction between the chains occurs, facilitating solvatisation. The same holds true for polysaccharides with a long linear chain and a high number of very short side chains like e.g. guaran. Charged polysaccharides in ordered conformation like pectin or carrageenan are also soluble because the charge on the polysaccharide and the tendency of the counter-ions to disperse in solvent favor dissolution. The viscosity of charged polysaccharide solutions is high and depends on the pH. Tables 3.3 and 3.4 show a selection of common homo-and heteropolysaccharides.

Starch is the principal nutritional reservoir in plants and in this way the major source of carbohydrates in human diet. It occurs in various amounts in seeds, fruits, leaves, tubers and bulbs from a few percent up to over 75% on cereal grains. Starch can be separated in two different components, in amylose and amylopectin, which vary in their relative amounts among the different starches. However, the majority of starches contain 15–35% amylose. Amylose, the unbranched type of starch consists of α-1,4-linked Glc residues (Figure 3.4a) with a degree of polymerization (DP) of 1'000–2'000 in cereal starches and up to 4'500 in potato starch corresponding to molecular weights of 150'000–750'000. Amylopectin which represents the branched form of starch, contains about one α-1,6-glycosidic linkage per 15–30 α-1,4-linkages distributed irregularly throughout the molecule (Figure 3.4b). It shows a molecular weight in the range of $1–20 \times 10^7$. In neutral aqueous solution the normal conformation of amylose is random coil. The presence of complexing agents in the solution induces formation of a helical structure consisting of about 6 D-glucosyl residues per helical turn [4]. This helical conformation gives rise to the characteristic blue color of amylose-iodine complexes and is responsible for the complex formation with fats and polar organic solvents. For amylopectin also helical structures are assumed with the double helices being arranged parallel to each other. Amylose and amylopectin are rapidly hydrolyzed by α-amylase

Table 3.3 Structure and source of homopolysaccharides

Name	Linkage	Source
D-Fructans		
Inulin	(2,1)-β-D- linear	dahlias, Jerusalem artichokes
Levan	(2,6)-β-D-, (2,1)-β-D- branched	plants and bacteria
D-Galactans		
Carrageenan	(1,3)-β-D-, (1,4)-α-D- linear	red seaweeds
	(1,4)-β-D- linear	plant pectic substances
D-Galacturonans		
Pectic acid	(1,4)-α-D- linear	plant pectic substance
D-Glucans		
Laminaran	(1,3)-β-D- linear	brown seaweeds, plants
Scleroglucan	(1,3)-β-D-, (1,6)-β-D- branched	fungi
Lichenan	(1,3)-β-D-, (1,4)-β-D- linear	iceland moss, cereal grains
Amylose	(1,4)-α-D- linear	plants
Amylopectin	(1,4)-α-D-, (1,6)-α-D- linear	plants
Glycogen	(1,4)-α-D-, (1,6)-α-D- linear	animals
Cellulose	(1,4)-β-D- linear	plant cell-walls
Pullulan	(1,4)-α-D-, (1,6)-α-D- linear	fungi
Dextran	(1,6)-β-D- linear	bacteria
2-Amino-2-deoxy-D-glucans		
Chitin	(1,4)-β-D- linear	crab and lobster shells, fungi
D-Mannans		
	(1,4)-β-D- linear	seaweed, plants
	(1,2)-α-D-, (1,6)-α-D- branched	yeasts
D-Xylans		
	(1,4)-β-D- linear	plant cell walls
Rhodymenan	(1,3)-β-D- linear	green seaweed

which hydrolyzes internal α-1,4-glycosidic linkages to yield maltose, maltotriose and α-dextrins in case of amylopectin. α-Dextrins are made up of several glucose units joined by an α-1,6-linkage in addition to the α-1,4-linkages. In contrast to α-amylase the β-amylase hydrolyzes the two starch components into maltose by sequential removal of disaccharide units from the non reducing end.

Table 3.4 Structure and source of heteropolysaccharides

Name	Constituent monosaccharides and chain type	Source
Agarose	DL-galactose, linear	red seaweeds
Alginic acid	L-guluronic acid, D-mannuronic acid, linear	bacteria, brown seaweeds
Gum arabicum	L-arabinose, L-rhamnose, D-galactose, D-glucuronic acid, branched	acacia trees
Gum tragacanth	D-galacturonic acid, D-xylose, L-fucose, D-galactose, branched	plant gum from *Astragalus spp.*
Guaran	D-mannose, L-galactose, branched	seeds from *Cyanopsis tetragonolaba*
Carubin	D-mannose, L-galactose, branched	seeds from *Ceratonia siliquia*
Pectin	D-galacturonic acid, L-rhamnose, D-galactose, L-arabinose, D-xylose, branched	plants
Xanthan	D-glucose, D-mannose, branched	bacteria

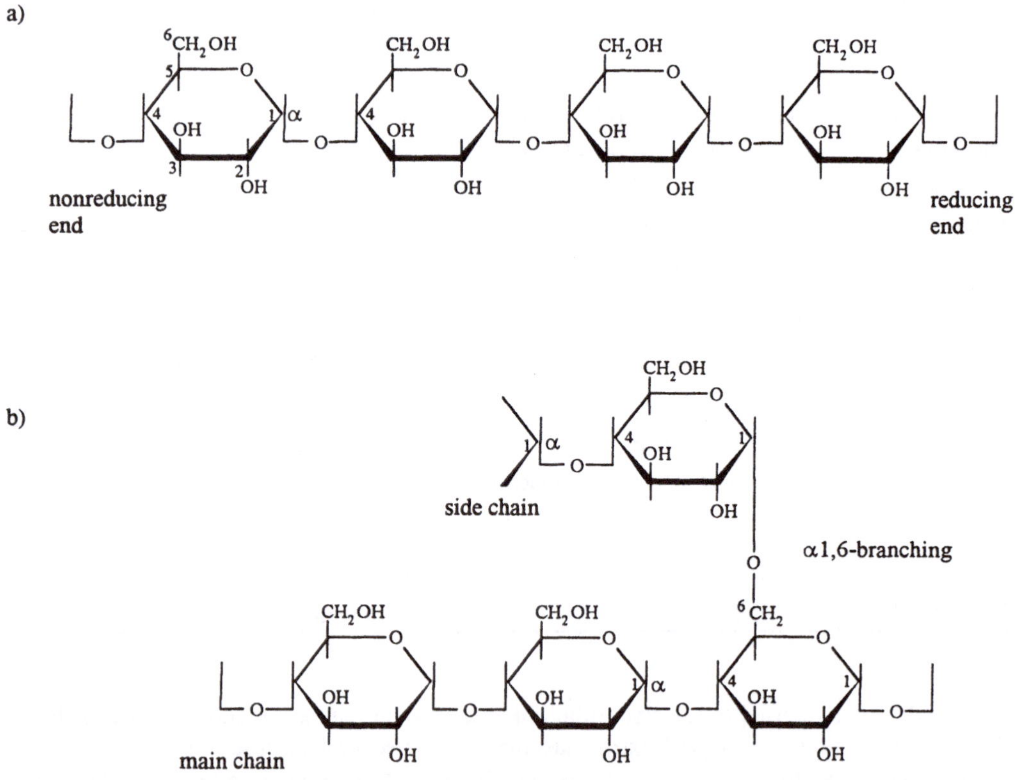

a)

b)

Figure 3.4 Structure of a) amylose and b) amylopectin.

Cellulose which represents the second major polysaccharide in plants serves as a structural element rather than as a nutrient. In fact, cellulose is the most abundant organic substance found in nature. Around 10^{15} kg are synthesized and degraded on earth each year [2]. It occurs in almost pure form (98%) in cotton fibers. Cellulose is a linear polymer consisting of β-1,4-linked Glc residues. The β-conformation allows cellulose to form very long straight chains, where each glucose residue is related to the next by rotation of 180°. The cellulose chains show the tendency to build microfibrils through inter- and intramolecular hydrogen bonding resulting in the formation of highly ordered structures with a high tensile strength. These ordered structures also cause the insolubility of cellulose in most solvents. Since mammals do not have cellulases they cannot digest vegetable fibers and wood.

The plant gums or exudate gums are a group of polysaccharides containing hexuronic acids in salt forms and a number of neutral monosaccharides which are often esterified in highly branched structures. These gums which may be formed spontaneously or by cutting the bark or fruit, are secreted as viscous liquids which become hard nodules by dehydration to seal the site of the injury in order to provide protection from microorganisms. They find industrial application as thickening agents and emulsion stabilizers. The best known example of plant gums is gum arabic from various *Acacia* species. It comprises a mixture of gums containing chains of β-1,3-linked Gal residues to which chains consisting of L-Ara, L-Rha and GlcA residues are attached. Since gum arabic occurs in its salt form it exhibits very good solubility in aqueous solution. Another example of a plant gum is gum tragacanth which is structurally related to pectic acid. It consists of a polygalacturonic acid chain to which Xyl, Gal and L-Fuc residues are attached.

The pectins are a group of substances found in the primary cell walls and intercellular layers in land plants. For example apples and citrus fruits contain 15 and 30%, respectively, of pectin in their dry weight [4]. Their principal constituent is Gal. A group of related polysaccharides which contain galacturonic acid instead of galactose are termed pectic acids while if the uronic residues are present as methyl esters they are called pectinic acids. Pectinic acids are easily extracted with water and show very good gelling properties which are used commercially in fruit jellies. In contrast, the pectic acids, usually present as their calcium salts, are less soluble. The most common pectic substances are heteropolysaccharides containing both acidic and neutral sugars, the latter in various proportions (10–25%). Among the neutral sugars only L-Rha was found to interrupt the polygalacturonic acid chain whilst the others like L-Ara, Gal and L-Xyl are attached in the side chains.

Like starch in plants, glycogen is the principal reserve polysaccharide in animals and for the most part stored in the liver. It is a very large molecule consisting of α-1,4-linked glucose chains and branches formed of α-1,6-linkages. The branches serve to increase the solubility of the molecule. Due to its high molecular weight of 10^7–10^8 and its short chain length of only 10–12 glucose units between branches, glycogen builds a highly branched tree like structure similar to that of amylopectin.

A number of specific polysaccharides is related to microorganisms where they occur as integral part of the cell wall, as capsules surrounding the cell or as extracellular polysaccharides which are formed in the culture media. The cell wall polysaccharides comprise the techoic acids which represent a group of phosphate containing polymers found in grampositive bacteria and the mureins which consist of polysaccharide chains in which individual residues carry an amino acid, cross-linked by peptide bridges [7]. The characteristic feature of the mureins are the amino-type monosaccharides GlcNAc and N-acetylmuramic acid (2-acetamido-3-O-(1-carboxyethyl-2-deoxy-D-glucose).

A special group of polysaccharides located on the outer surface of gram-negative bacteria, forming an integral part of the outer membrane structure (capsular polysaccharides) are the lipopolysaccharides (LPS). These complex membrane components, also referred to as endotoxins, are considered essential for the maintenance, growth and reproduction of the cell. Additionally, LPS are the main antigens of gram-negative bacteria and can induce an immunological response from the host immune system which arises from the non reducing end groups of the LPS polysaccharide part [7, 8]. Chemically, LPS consist of two parts, of a hydrophilic polysaccharide and a hydrophobic glycolipid, as depicted in Figure 3.5. The latter is referred to as lipid A and consists of a glucosamine disaccharide containing N-and O-linked fatty acids. The polysaccharide part comprises the O-specific chain and the core oligosaccharide which is relatively conserved in its composition and structure compared to the O-specific chain. The inner core oligosaccharides of the core structure include unique characteristic features of LPS such as 3-deoxy-β-D-manno-octopyranulosonic acid (KDO) and L-glycero-β-D-manno-heptopyranose (Hep). For structures see Figure 3.5. The outer

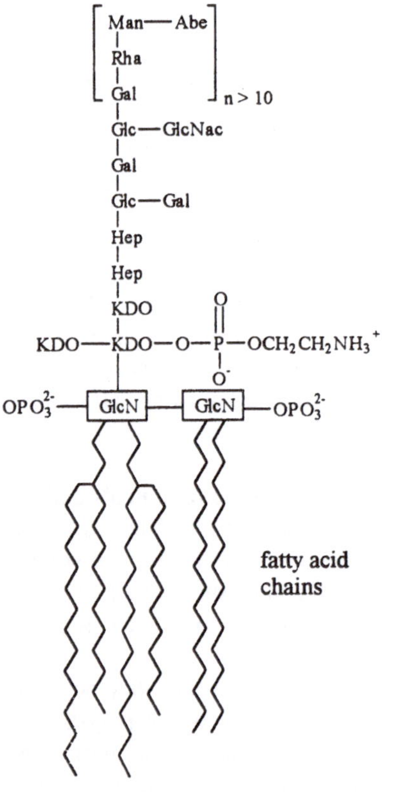

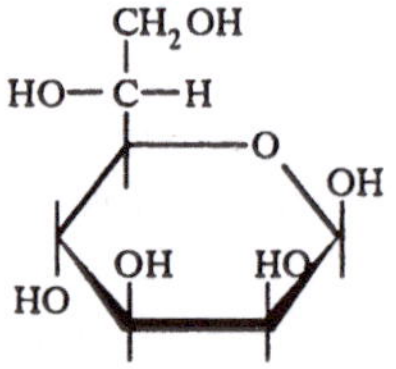

L-glycero-β-D-manno-heptopyranose (Hep)

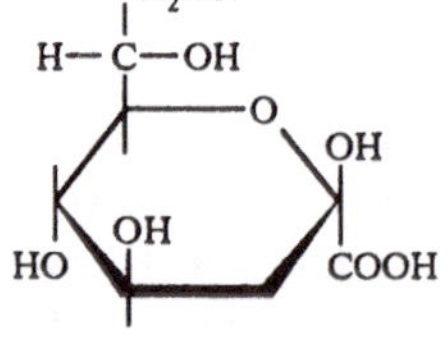

3-Deoxy-β-D-manno-octopyranulosonic acid (KDO)

Figure 3.5 Outer membrane lipopolysaccharide of *Salmonella typhimurium*. Abe = Abequose core consists of hexoses, primarily Glc, Gal and GlcNAc, all in their pyranose form with α-anomeric configuration [9]. According to the nature of the O-specific chain moiety, which is composed of repeating oligosaccharide units, gram-negative bacteria are classified into bacterial serotypes. Each bacterial serotype produces an unique LPS which is characterized by a specific composition and structure of the O-chain and which leads to an individual O-antigenicity [10].

Bacteria produce a number of polysaccharides with the most important being the dextrans which are used as plasma substituents and, in a modified form, as molecular sieves. The main chain of dextrans consists of α-1,6-linked glucose residues with differing amounts of α-1,3- and α-1,4-linked branching points, depending on the bacteria.

3.4 Glycoproteins

Many macromolecules such as integral membrane proteins, serum and plasma proteins, antibodies or clotting factors are known to be glycosylated. These glycoproteins consist of a protein chain of about 200 amino acids to which hetero-oligosaccharide chains are covalently attached. The oligosaccharide chains are usually branched and can contain neutral monosaccharides (Glc, Gal, Man, L-Fuc), basic monosaccharides (GlcNAc, GalNAc) and the nine-carbon sugar neuraminic acid which can be N-acetylated or N-glycolylated (Figure 3.6).

Since glycosylation of proteins is a posttranslational modification which depends on the presence of certain enzymes and no template is included in the biosynthesis of the oligosaccharide chains, structural microheterogeneity is the main characteristic of the carbohydrate part in glycoproteins. Oligosaccharides are attached to proteins in one of three ways:

(1) through an N-glycosidic bond to the side chain of an asparagine residue

(2) through an O-glycosidic bond to the side chain of serine, threonine, hydroxylysine or hydroxyproline,

(3) as part of an glycosylphosphatidylinositol anchor [11].

Usually, each glycosylation site in an eukariotic glycoprotein is associated with several different oligosaccharide structures, a phenomenon referred to as N- or O-glycosylation site microheterogeneity. These oligosaccharide structures are characterized by the type and se-

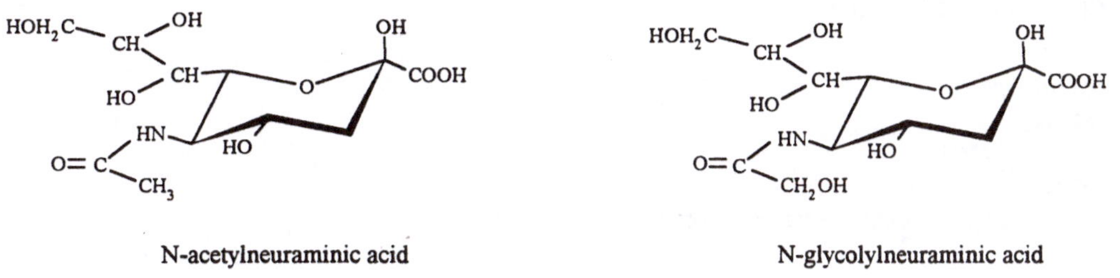

N-acetylneuraminic acid   N-glycolylneuraminic acid

Figure 3.6 Structure of N-acetylneuraminic acid and N-glycolylneuraminic acid

quence of monosaccharides, the position and anomerity of the glycosidic linkage and the branching pattern [12]. Further, each potential glycosylation site in a given glycoprotein population is fully, partially, or not at all occupied. In fact, an eukaryotic glycoprotein is not isolated as a single structural entity, but rather as a set of glycosylated variants of a common polypeptide, which are referred to as glycoforms [11, 13]. The particular glycosylation pattern of a protein is not random and uncontrolled and reflects the balance of all activities of that glycoprotein in a particular physiological state.

All N-linked oligosaccharides share the same pentasaccharide core structure:
Manα-1,6-(Manα-1,3)-Manβ-1,4-GlcNAcβ-1,4-GlcNAc. This core element is extended through attachment of a number of monosaccharides in different branching patterns with usually two, three or four branches or antennaries, respectively. Based on compositional similarities the oligosaccharides are further classified in three subgroups as shown in Figure 3.7. *High mannose type glycans* contain only α-mannosyl residues in addition to the pentasaccharide core. In contrast the *complex type glycans* contain other monosaccharide residues in the outer chains such as GlcNAc, Gal or NeuAc, but no additional Man-residues. The presence or absence of a α-Fuc attached to the C-6 position of the proximal GlcNAc and the GlcNAc attached to the C-4 position of the β-Man in the core structure (bisecting GlcNAc) contributes to the structural variety of the complex type glycans. The third group is called the *hybrid type* since these oligosaccharides combine the features of both the complex and the high mannose type glycans. They contain Man-residues in the α-1,6 arm and other than Man-residues in the α-1,3 branch of the core. A fourth group of N-glycans are the poly-

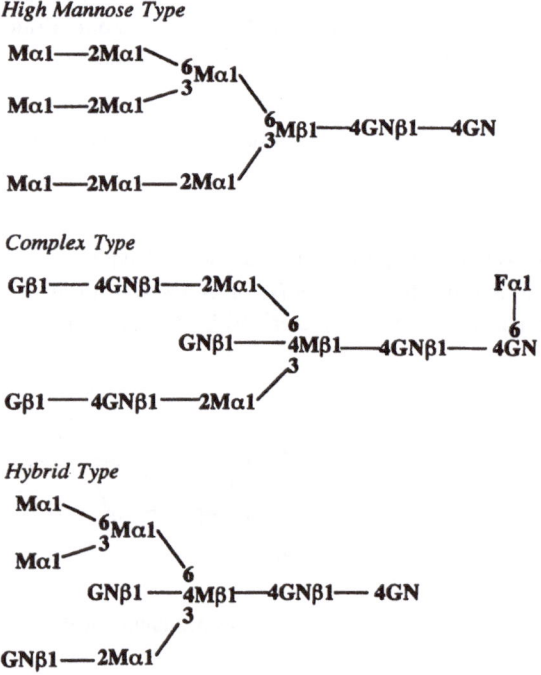

Figure 3.7 Structures of N-linked glycans; M = Man, GN = GlcNAc, G = Gal, F = Fuc

N-acetyllactosamines containing repeating units of (Galβ-1,4-GlcNAcβ-1,3-) attached to the core [14-16].

Unlike the N-linked glycans O-glycans have fewer structural rules. They can be classified into at least four groups according to the different core structures shown in Figure 3.8. These cores can be elongated by addition of Gal in β-1,3- and β-1,4-linkages and GlcNAc in β-1,3- and β-1,6-linkages. The termini are formed through addition of Gal, GalNAc or L-Fuc. In addition, single monosaccharide residues such as GlcNAc or L-Fuc may be O-linked to the peptide backbone [15, 16].

core 1

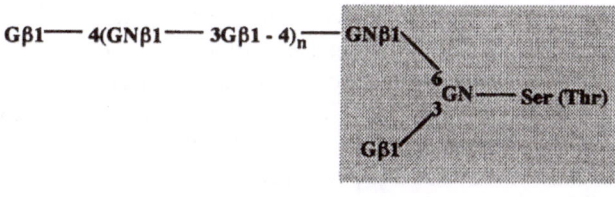

core 2

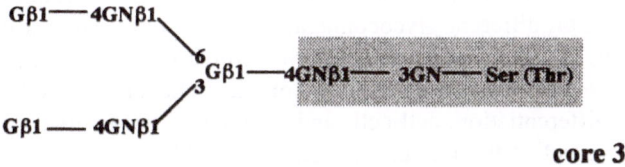

core 3

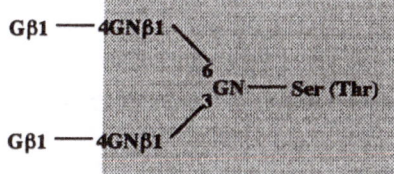

core 4

Figure 3.8 Core structures typically found in O-linked glycans. The cores are enclosed in the tinted boxes. GN = GlcNAc, G = Gal, Ser = serine, Thr = threonine.

The oligosaccharide moieties can have considerable effects on the bioactivity, the solubility and the stability of proteins, since they are involved in correct folding, and the susceptibility of glycoproteins to proteolysis. Further, these moieties are important in determining their pharmacokinetic profile and their clearance rate. Recognition of oligosaccharide sequences by certain receptors, such as the hepatic Gal/GalNAc-receptor (asialoglycoprotein-receptor) which is located on liver cells, can result in removal of the glycoconjugate or even the whole cell from the circulation. Many of the recombinant glycoproteins such as immunoglobulins and peptide hormones contain glycans with terminal sialic acid residues which are removed by sialydases on the surface of blood vessels. The exposed Gal residues are detected by the Gal/GalNAc-receptor leading to the removal of the trimmed glycoprotein from the blood. The deficiency of terminal sialic acid or galactosyl residues in glycan structures can also cause certain diseases, such as rheumatoid arthritis, due to an increased clearance rate of the specific glycoprotein, in this case the immunoglobulin G.

Intercellular recognition and adhesion as well as control of cell differentiation and cell development are mediated by interactions of carbohydrates located on the cell surface. Cells communicate with their environment by means of this sugar 'coating'. One example for cell-cell interactions caused by complex oligosaccharides is that of the Selectin receptor proteins which mediate the adhesion of leukocytes to endothelial cells and the recognition of leukocytes by stimulated or wounded endothelium. Complex oligosaccharides can also act as highly specific receptors for a variety of viruses and bacteria, for plant and bacterial toxins as well as for hormones and antigens for autoimmune reactions. In most of these instances an exquisite specificity for the sequence of the oligosaccharide is involved. The addition of specific monosaccharides to the oligosaccharide can mask the special sequence and prevent its recognition by microbial toxins or antibodies. Intracellular targeting of proteins may also depend on the glycan structure associated with the protein as was demonstrated for lyosomal enzymes.

Altogether, glycoproteins can have many different functions (Table 3.5). Expression of specific types of glycosylation on different glycoconjugates in different tissues at different times during development imply that these structures have diverse roles in the same organism. Glycoprotein glycans are essential in a variety of biological phenomena, both normal, such as embryogenesis, cell differentiation, cell-cell- and receptor-ligand interaction, targeting of cells to tissues and control of the immune system, and pathological, such as tumor progression, leukemia or viral and bacterial infectivity [17]. Varki [18], Rademacher et al. [13] and Paulson [14] reviewed comprehensively the various aspects of the biosynthesis and the biological role of oligosaccharides in glycoproteins.

3.5 Glycosaminoglycans and Proteoglycans

Unlike glycoproteins where the protein part is the predominant part, in proteoglycans the carbohydrates usually dominate with 100–200 monosaccharide residues per molecule. Proteoglycans consist of a protein backbone to which linear carbohydrate chains of alternating disaccharide units, the glycosaminoglycans, are covalently attached. These disaccharide units comprise acidic monosaccharides such as GlcA and IdoA and basic monosaccharides

Table 3.5 Function of some glycoproteins

Presumed function	Glycoprotein	Presumed function	Glycoprotein
Enzyme	Acetyl cholinesterase Ficin Pancreatic ribonuclease Yeast invertase Yeast acid phosphatase Porcine α-amylase	Receptor	Asialoglycoprotein receptor Thyrotropin receptor PDGF receptor Transferrin receptor Insulin and Insulin-like growth factor receptor
Hormone	Erythropoietin Follicle-stimulating hormone Human chorionic gonadotrophin Luteinizing hormone Thyroglobulin	Protective	Immunoglobulin Interferon Fibrinogen Mucins
Transport	Ceruloplasmin Transferrin	Serum proteins	α$_1$-Acid glycoprotein Fetuin
Coagulation	Tissue-type plasminogen activator Plasminogen Antithrombin IIIb	Structural	Bacterial cell wall Collagen Extensin (plant cell wall)
Food. Reserve	Casein Endosperm glycoproteins Ovalbumin		

such as GlcNAc and GalNAc. The basic units are usually N-sulfated and the acidic units O-sulfated. There are four general classes of glycosaminoglycans: (1) hyaluronic acid, (2) chondroitin-sulfate and dermatan-sulfate, (3) keratan sulfate, and (4) heparan sulfate and heparin. The structural features of the glycosaminoglycan disaccharide units are given in Table 3.6 [19].

Hyaluronic acid is the simplest glycosaminoglycan, with GlcA and GlcNAc residues linked through β-1,4- and β-1,3-linkages. Since hyaluronic acid is not synthesized covalently bound to a protein like the other glycosaminoglycans, it is not classified as a proteoglycan [19]. Chondroitin sulfate has the same backbone structure as hyaluronic acids except that the GlcNAc residues are replaced by GalNAc residues. Sulfate esters are attached to the C-4 and

Table 3.6 Classification of glycosaminoglycans

	Type	Linkgage	Sulfate residues
(1)	Hyaluronic acid	-4-D-GlcAβ1-3-D-GlcNAcβ1-	no
(2)	Chrondoitin-4-sulfate -6-sulfate	-4-D-GlcAβ1-3-D-GalNAcβ1-	4-position of GalNAc 6-position of GalNAc
	Dermatan sulfate	-4-L-IdoAα1-3-D-GalNAcβ1-	4- or 6-position of GalNAc
(3)	Keratan sulfate	-3-D-Galβ1-4-D-GlcNAcβ1-	6-position of Gal and GlcNAc
(4)	Heparan sulfate Heparin	-4-D-GlcAβ1-4-D-GlcNAcα1- -4-L-IdoAα1-4-D-GlcNAcα1-	2-and 6-position of GlcNAc 2-position of Ido, 2-and 6-position of GlcNAc

C-6 positions of the GalNAc. In dermatan sulfate only the GlcA residues of the chondroitin sulfate are epimerized into IdoA residues. Like hyaluronic acid and chondroitin sulfate, keratan sulfate contains alternating β-1,4- and β-1,3-linkages, but with GlcNAc in the hexuronic acid position and Gal in the hexosamine position. Sulfate esters are generally attached to the C-6 position on one or both of the sugar units. The repeating disaccharide structure of heparan sulfate and heparin is distinctly different from that of the other glycosaminoglycans. The linkage of GlcNAc to GlcA is α-1,4 instead of β-1,4 and that of GlcA and GlcNAc is β-1,4 instead of β-1,3.

Glycosaminoglycans are linked to the protein backbone via a glycopeptide linkage where the reducing end of the terminal monosaccharide is linked to the side chain of an amino acid residue which usually is a serine. The linkage involves monosaccharide units different from those of the main polysaccharide chain. For keratan sulfate the linkage occurs through oligosaccharide structures which are related to O- and N-linked oligosaccharides present in glycoproteins while for the other glycosaminoglycans it occurs through a characteristic tetrasaccharide [19].

Proteoglycans and glycosaminoglycans play a very significant role in animal organisms since they participate considerably in the formation of connective tissue which supports and binds organs and bones. They also form collagen fibers, cell membranes, extracellular fibrils and cartilage. Connective tissue consists of collagen fibrils and proteoglycans together with aqueous compounds. In the tissue the protein chains and the collagen fibers lie side by side whilst the glycosaminoglycans, e.g. chondroitin sulfate, interact with the collagen forming a three dimensional matrix. This matrix resists disruption and separation, encloses organs and provides a buffer to external forces. Aggregating proteoglycans, called aggrecan, form enormous macromolecules with hyaluronic acid. To this complex, which represents the major cartilage proteoglycan, about 100 chondroitin-sulfate chains and a smaller number of keratan-sulfate chains are attached.

Other than maintaining the overall structure of tissues, glycosaminoglycans are involved in wound healing (heparin), the control of urine concentration in the kidney, the storage and release of biogenic amines, the maintenance of stable transport media, the maintenance of the cornea in a transparent form and the control of the hydration properties of the cornea [7].

For several years heparin has been used as a clinical anticoagulant. By stimulating the activity of serine proteinase inhibitors, heparin inhibits serine proteases thus blocking the coagulation cascade. Through its action on platelets and on the endothelium, heparin also serves as an antithrombotic agent [20]. However, the application of heparin is not limited to the therapeutic treatment of established thrombi. It was also found to be of major use in the prevention of postsurgical thrombosis. The prophylactic use of heparin allows the application of lower dosages thus preventing persistent hypocoagulation which often is associated with its therapeutic use [21]. A second glycosaminoglycan under clinical investigation as a new antithrombotic agent is dermatan sulfate. Unlike heparin dermatan sulfate acts more selectively by exhibiting only a weak anticoagulant effect [20].

Since proteoglycans occur widely distributed in the human body, defects in many diseases will be related in some way to them. Due to the relationship between proteoglycans and other tissues, the effect on one proteoglycan may have far-reaching consequences on the maintenance of production and function of intact tissue. So far the largest group of proteoglycan related diseases is a group of hereditary diseases characterized by excessive urinary excretion of glycosaminoglycan material. These diseases which are called 'genetic hyperglycosaminoglycanuria' involve complex storage disorders in which both glycosaminoglycans and other materials are affected and accumulated in tissues. The primary defect is an enzyme deficiency resulting in faulty breakdown of glycosaminoglycans. Many of the conditions of genetic hyperglycosaminoglycanuria are manifested in childhood causing extensive deformity and mental retardation [7].

3.6 Glycolipids

Glycolipids represent a diffuse class of molecular species covering a wide range of structural types. More than 200 glycosphingolipids, differing in their glycan moieties, have been isolated from mammalian tissues [22]. They are classified in neutral glycosphingolipids, sulfatoglycosphingolipids, fucoglycosphingolipids and gangliosides. Usually glycolipids are associated with membranes by non covalent bonds, with the hydrophobic region buried in the outer membrane lipids and the carbohydrate region extending in the aqueous phase. Most animal glycolipids are derivatives of the aminoalcohol sphingosine to which long chain fatty acids (C16–C24) are attached through amide linkages and which are called ceramides (Figure 3.9). They contain carbohydrate moieties with one to seven residues in a linear or branched structure linked to the terminal hydroxyl group of the sphingosine base. The most common components are Glc, Gal, GlcNAc and GalNAc. Figure 3.10 shows a selection of some major saccharide cores found in animal glycosphingolipids.

The simplest glycolipids found in most mammalian tissues and in specially high concentrations in the central nervous system, are the monoglycosylceramides or cerebrosides. Brain cerebrosides contain for the most part D-galactosyl groups whilst those from serum contain D-glucosyl groups. Diglycosylceramides such as lactosyl ceramide are also widely distributed. They are precursors for more complex glycosylceramides and gangliosides [7].

Gangliosides are NeuAc containing glycosphingolipids with one or more NeuAc residues linked to the glycosylceramide as single residue or as disaccharide residue side chain. The nomenclature for these gangliosides consists of the prefix G (for ganglioside) with the subscripts M (mono), D (di) or T (tri) to denote the number of N-acetylneuraminic acid residues (Figure 3.9) [7, 23]. Stearic acid is the predominant fatty acid (85–95%) in gangliosides found commonly in spleen, liver and kidney with the gray matter of brain being the major site of occurrence. Gangliosides are assumed to play important roles in the regulation of biological processes such as cell growth and differentiation. In addition, they are also potential drugs. For example the ganglioside G_{M1} was shown to have a therapeutic effect on Alzheimer's disease. Other gangliosides are overexpressed as antigens in several cancer cells, indicating the potential applicability in immunotherapy [24].

It is assumed that glycolipids function as carriers to transport carbohydrate moieties across cell membranes or as modifiers which alter the physiological properties of the membranes. They are known to be involved in the biosynthesis of glycoproteins and complex polysaccharides with the lipid part acting as a carrier for the growing carbohydrate moiety [7]. Additionally, they provide protection against toxins such as the cholera or the tetanus toxin, through binding of these toxins to the carbohydrate moiety of the glycolipid. Involvement of glycolipids in diseases is caused by genetic defects resulting in enzyme deficiencies

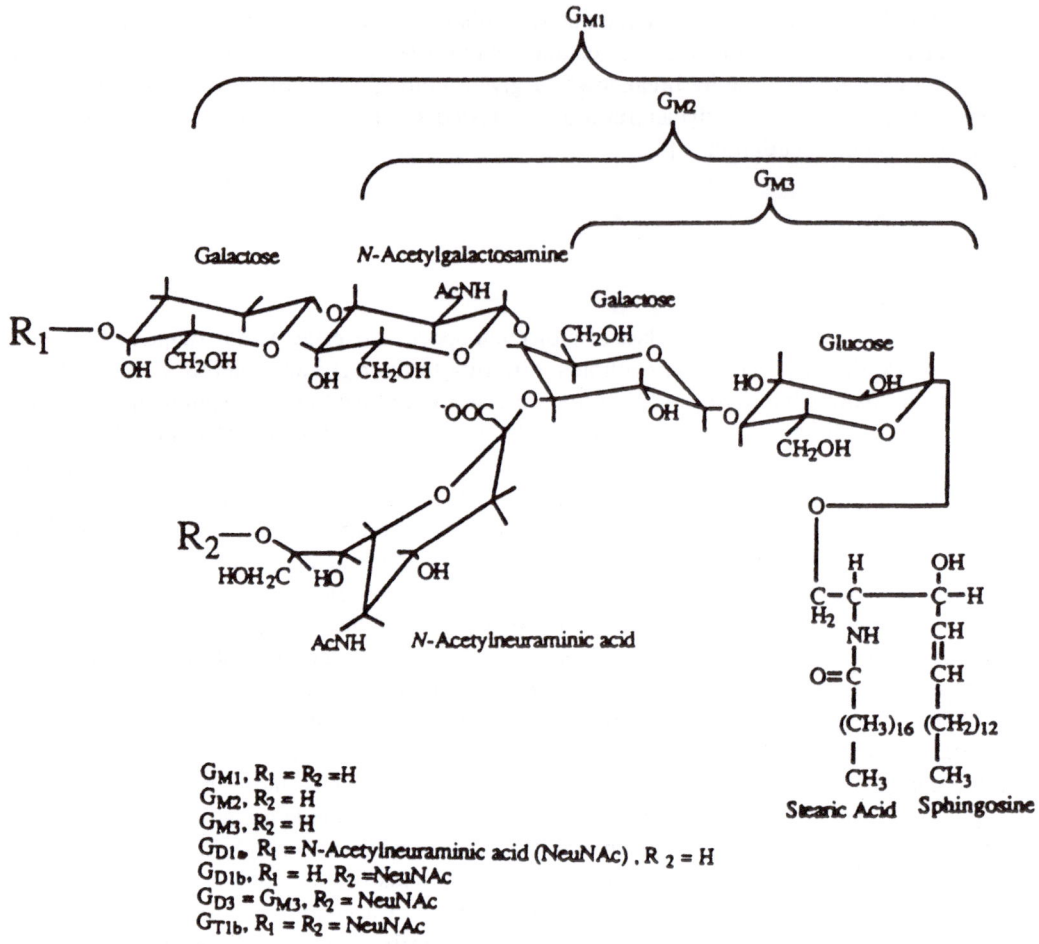

Figure 3.9 Ganglioside structures (from [23], with permission)

and an accumulation of the distinct lipid or glycolipid. This leads to a disease known as liposidosis. The liposidoses belong to a group of rare inborn defects in metabolism which result in mental retardation and are frequently fatal.

Gangliotetraose (ganglio-series)	Galβ1 – 3GalNAcβ1 – 4Galβ1 – 4Glc
Globotetraose (globo-series)	GalNAcβ1 – 3Galα1 – 4Galβ1 – 4Glc
Isoglobotetraose (isoglobo-series)	GalNAcβ1 – 3Galα1 – 3Galβ1 – 4Glc
Lacto-N-tetraose (lacto-series)	Galβ1 – 3GlcNAcβ1 – 3Galβ1 – 4Glc

Figure 3.10 Major saccharide cores found in animal glycosphingolipids

References

1. J.D. Olechno, and K.J. Ulfelder, "Carbohydrate analysis by capillary electrophoresis", in "Handbook of capillary electrophoresis. A practical approach", (ed. J.P. Landers), CRC Press, Boca Raton **1994**.

2. A.L. Lehninger, D.L. Nelson, and M.M. Cox, "Prinzipien der Biochemie", 2nd ed., Spektrum Akademischer Verlag, Heidelberg **1994**.

3. L. Stryer, "Biochemistry", 4th ed., W.H. Freeman & Co **1995**.

4. H.-D. Belitz, and W. Grosch, "Lehrbuch der Lebensmittelchemie", 3rd ed. Springer-Verlag, Berlin **1987**.

5. R.T. Morrison, "Organic Chemistry", 6th ed., Prentice Hall **1992**.

6. I.A. Streitwieser, C.H. Heathcock, and E.M. Kosower, "Introduction to Organic Chemistry", 4th ed., Prentice Hall **1992**.

7. J.F. Kennedey, and C.A. White, "Bioactive carbohydrates in chemistry, biochemistry and biology", Ellis Horwood Ltd., John Wiley & Sons, Chichester, New York **1983**.

8. J. Kelly, H. Masoud, M.B. Perry, J.C. Richards, and P. Thibault, "Separation and characterization of O-deacylated lipooligosaccharides and glycans derived from Moraxella catarrhalis using capillary electrophoresis-electrospray mass spectrometry and tandem mass spectrometry", *Anal. Biochem.*, *233* **1996** 15-30.

9. K.P. Bateman, J.H. Banoub, and P. Thibault, "Probing the microheterogeneity of O-specific chains from *Yersinia ruckeri* using capillary zone electrophoresis/electrospray mass spectrometry", *Electrophoresis*, *17* **1996** 1818-1828.

10. O. Lüderitz, M.A. Freudenberg, C. Galanos, V. Lehmann, E.T. Rietschel, and D.H. Shaw, "Lipopolysaccharides of gram-negative bacteria", *Curr. Top. Membr. and Transp.*, *17* **1982** 79-151.

11. R.A. Dwek, C.J. Edge, D.J. Harvey, M.R. Wormald, and R.B. Parekh, "Analysis of glycoprotein-associated oligosaccharides", *Annu. Rev. Biochem.*, *62* **1993** 65-100.

12. M.W. Spellman, "Carbohydrate characterization of recombinant glycoproteins of pharmaceutical interest", *Anal. Chem.*, *62* **1990** 1714-1722.

13. T.W. Rademacher, R.B. Parekh, and R.A. Dwek, "Glycobiology", *Ann. Rev. Biochem.*, *57* **1988** 785-838.

14. J.C. Paulson, "Glycoproteins: what are the sugar chains for?", *TIBS*, *14* **1989** 272-276.

15. R. Kornfeld, and S. Kornfeld, "Comparative aspects of glycoprotein structure", in "Annual Review of Biochemistry" (ed. E.E. Snell), Annual Review Inc., Palo Alto **1976**.

16. Oxford Glycosystems, "Tools for glycobiology", catalog **1994**.

17. H. Schachter, "Branching of N- and O-glycans: biosynthetic controls and functions", *Trends Glycosci. Glycotechnol.*, *17* **1992** 241-250.

18. A. Varki, "Biological role of oligosaccharides: all of the theories are correct", *Glycobiology*, *3* **1993** 97-130.

19. V.C. Hascall, A. Calabro, R.J. Midura, and M. Yanagishita, "Isolation and characterization of proteoglycans", in "Guide to techniques in glycobiology", (ed. W.J. Lennarz and G.W. Hart), Academic Press, San Diego **1994**.

20. R.J. Linhardt, U.R. Desai, J. Liu, A. Pervin, D. Hoppensteadt, and J. Fareed, "Low molecular weight dermatan sulfate as an antithrombotic agent", *Biochem. Pharmacology*, *47* **1994** 1241-1252.

21. J.B.L. Damm, G.T. Overklift, B.W.M. Vermeulen, C.F. Fluitsma, and G.W.K. van Dedem, "Separation of natural and synthetic heparin fragments by high-performance capillary electrophoresis", *J. Chromatogr.*, *608* **1992** 297-309.

22. C.L.M. Stults, C.C. Sweely, and B.A. Macher, "Glycosphingolipids: Structure, biological source, and properties", in "Methods in Enzymology", (ed. V. Ginburg), Academic Press, San Diego **1989**.

23. Y. Mechref, G.K. Ostrander, and Z. El Rassi, "Capillary electrophoresis of carboxylated carbohydrates I. Selective precolumn derivatization of gangliosides with UV absorbing and fluorescent tags", *J. Chromatogr. A*, *695* **1995** 83-95.

24. Y.S. Yoo, and Y.S. Kim, "Separation of gangliosides using cyclodextrin in capillary zone electrophoresis", *J. Chromatogr. A*, *652* **1993** 431-439.

4 Separation and detection of carbohydrates in capillary electrophoresis

At first sight, the analysis of carbohydrates with electrophoresis seems to be a mismatch. As discussed in Chapter 2, electrophoretic separations rely on differential movement of the sample components in an electrical field. Apart from MECC and CEC, which are based on differences in partition and chromatographic interactions, all CE modes require charged solutes. Except for some sulfonated, acetylated and phosphorylated carbohydrates, carbohydrates are in general neutral compounds and consequently exhibit no movement in an electrical field.

In addition, the commercial CE instrumentation is equipped with on-line spectral detection methods (UV and LIF). Carbohydrates generally lacking functional groups apart from multiple hydroxyl groups, which define them as a chemical class, have extremely low extinction coefficients. The molar extinction coefficient in the low UV at 195 nm for monosaccharides is only 2–5 l mol^{-1} cm^{-1} for aldoses and 12 l mol^{-1} cm^{-1} for fructose. This compares to an extinction coefficient of several hundreds to thousands for multi-ring compounds with conjugated functional groups. The UV concentration detection limits of 10^{-5}–10^{-6} M quoted in Chapter 2 are based on extinction coefficients of 1'000–10'000 l mol^{-1} cm^{-1}. Given that carbohydrates absorb very little UV light, they are also not fluorescent, excluding also LIF as a detection option for natural saccharides.

For these reasons, one is forced to somehow attach a charge to the carbohydrates of interest and simultaneously develop a detection scheme to meet the sensitivity requirements for a particular analysis problem. Several strategies how to attach a charge to the neutral carbohydrates have been developed and will be discussed below:

- complexation with borate
- ionization at high pH > 11.5
- derivatization with a charged label

In terms of detection, also a couple of options emerged, among them the two commercial options UV and LIF detection and home built electrochemical detection schemes. The main purpose of this chapter is to assess the different approaches taken to address the problem of high performance separation and sensitive detection of carbohydrates and glycoconjugates in CE and to outline some general trends.

4.1 Separation and detection of non-derivatized carbohydrates

4.1.1 Electrophoresis of native carbohydrates with direct UV detection

4.1.1.1 CZE and MECC separation conditions

Despite the fact that most carbohydrates are neutral compounds, some do have an intrinsic charge and are therefore suitable for electrophoretic analysis. In addition, the charged functional group also increases the extinction coefficient to allow UV detection using CZE conditions. The acidic disaccharides derived from the glycosylaminoglycans chondroitin, heparin and hyaluronic acid are among those charged carbohydrates. Because the enzymatic cleavage results in sulfated products they can consequently be analyzed in acidic buffer systems with low EOF. For example, a mixture of non-, mono-, di- and trisulfated disaccharides could be separated in a 20 mM phosphoric acid BGE at pH 3.48 [1]. In general charged solutes show an improved resolution under acidic conditions due to narrow peaks with close to Gaussian symmetry. In contrast to that, a series of heparin derived oligosaccharides exhibited a substantial loss in resolution in an acidic BGE, compared to the separation in the alkaline SDS/borate buffer. Non-sulfated glycosylaminoglycans do not migrate in acidic BGE as they carry no charge at low pH.

Separation of natural and synthetic LMW heparin fragments was achieved at low pH conditions with direct UV detection at 214 nm [2]. In this case quantitation of the electropherogram was not straightforward, since UV response of the synthetic precursors in the heparin preparations, containing strong UV absorbing substituents, was much higher than in the end-products. An accurate quality control of the pharmaceutical heparin preparations appears therefore difficult.

In a series of papers the group of Linhardt showed that unsaturated uronic acid residues, resulting from enzymatic cleavage of gylcosaminoglycans (see Chapter 5.5), allow UV detection at 232 nm at a micromolar level [1, 3, 4]. They used a basic MECC buffer with high EOF, composed of 50 mM SDS in a 10 mM sodium borate buffer, pH 8.8 to separate heparin and heparan sulfate derived disaccharides. The same buffer system allows the analysis of LMW heparin oligosaccharides with four to six carbohydrate residues with the migration velocity depending on their charge-to-mass ratio.

Glycoprotein derived sialooligosaccharides can be detected at 185 nm due to the presence of N-acetyl-groups in their sialic acid and hexosamine residues. However, a 185 nm detection requires a new deuterium lamp with high output and meticulously purified buffers. Oligosaccharides containing N-acetylneuraminic and N-glycoylneuraminic acid have been separated in a SDS/borate buffer, pH 9.6 (see Figure 5.24) [5].

4.1.1.2 CZE after on-column complexation with borate

It has long been known, that the tetrahydroxy borate ion $B(OH)_4^-$ can form complexes with polyhydroxy compounds [6]. Figure 4.1 displays the most frequently discussed mechanism of the borate complexation. At alkaline pH, boric acid acts not as proton donor but rather as a Lewis acid by accepting a hydroxyl ion to form $B(OH)_4^-$. The tetrahydroxy borate can form stable five and six-membered ring structures with cis-1,2 or cis-1,3 diols, either as mono-borate or di-borate complex.

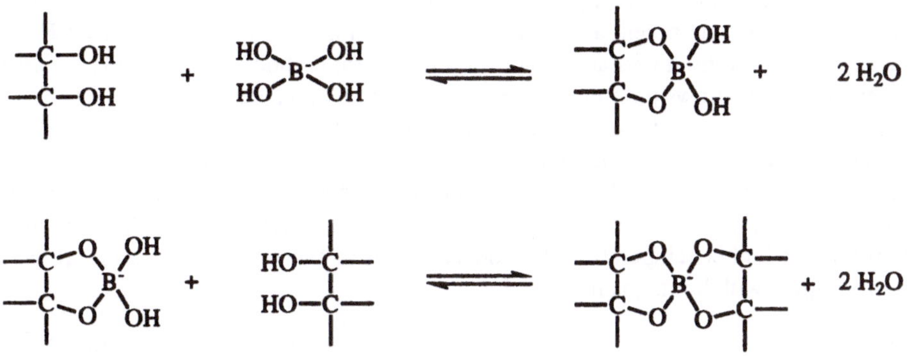

Figure 4.1 Complexation of carbohydrate polyols with borate

The formation of the mono-borate complex is instantaneously and the equilibrium towards complex formation. The stability of di-borate complex depends very much on the nature of the polyol. It is interesting to note that the actual concentration of the complexing ion tetrahydroxy borate depends on the salt used to make up the aqueous solution and the pH. Certainly, highly alkaline pH favors the formation of the tetrahydroxy borate ion. In addition, boron can form a multitude of stable polynuclear complexes with oxygen, such as the tetraborate ion and a number of polyborate ions. To ensure the maximum concentration of the tetrahydoxy borate ions possible, one should always use boric acid as buffer or electrolyte component and not sodium tetraborate or other polynuclear borate salts. Only at pH above 12, the tetraborate forms readily tetrahydroxy borate ions. In addition, the sodium will add to the buffer conductivity, especially at borate concentrations above 100 mM.

As shown in Figure 4.1, carbohydrates form in situ negatively charged complexes with borate. The fraction of the negative charge, effective for electrophoresis is determined by the equilibrium constant and therefore by the stability of the complex. According to the law of mass action the complex concentration increases with increasing borate concentration at constant carbohydrate concentration. Due to higher concentrations of borate ions in alkaline solutions the formation of sugar-borate complexes is favored at higher pH.

Furthermore, the stability of the sugar-borate complex depends strongly on the configuration, the number of hydroxyl groups and substituents in the carbohydrate molecule. Since borate tends to form a more stable complex with cis- than with trans-oriented pairs of hydroxyl groups on adjacent and alternate carbon atoms, small steric differences between closely related isomers are sufficient for a separation of those carbohydrates. Sterically, the most favorable configuration is a cis-oriented pair of hydroxyl groups at C_2 and C_4. Since for different carbohydrates the open chain as well as the annular forms can carry vicinal hydroxyl groups, it is assumed, that both forms are amenable to complexation.

The on-column complexation of native carbohydrates with borate allows direct UV detection at 195 nm as first demonstrated by Hoffstetter-Kuhn et al. [7]. Surprisingly, a two- to twenty fold increase in UV absorbance compared to non-borate complexed carbohydrates is observed. The data suggest that the complex formation favors the open chain structure of the carbohydrate. The overall sensitivity of on-column borate complexation for non-derivatized carbohydrates is in the order of 10^{-3} M. Although this approach may therefore not be attrac-

tive for carbohydrate analysis with high sensitivity requirements such as carbohydrate elucidation in glycoproteins, the potential of borate complexation for sugar analysis by CE was
clearly demonstrated. With the borate complexation, neutral carbohydrates acquire a partial
negative charge, which allows their migration and separation in an electrical field.

Separation efficiency, resolution and analysis time for the analysis of mono- and disaccharides could be improved at higher temperature. For example, glucose and xylose, which
co-eluted at 20°C after more than 40 minutes could be baseline resolved at 60°C within only
20 minutes (Figure 4.2). Both solutes have no vicinal cis-hydroxyl groups in their pyranose
forms, however, at higher temperature the concentration of the open chain form, which is
assumed to complex with borate, is increased.

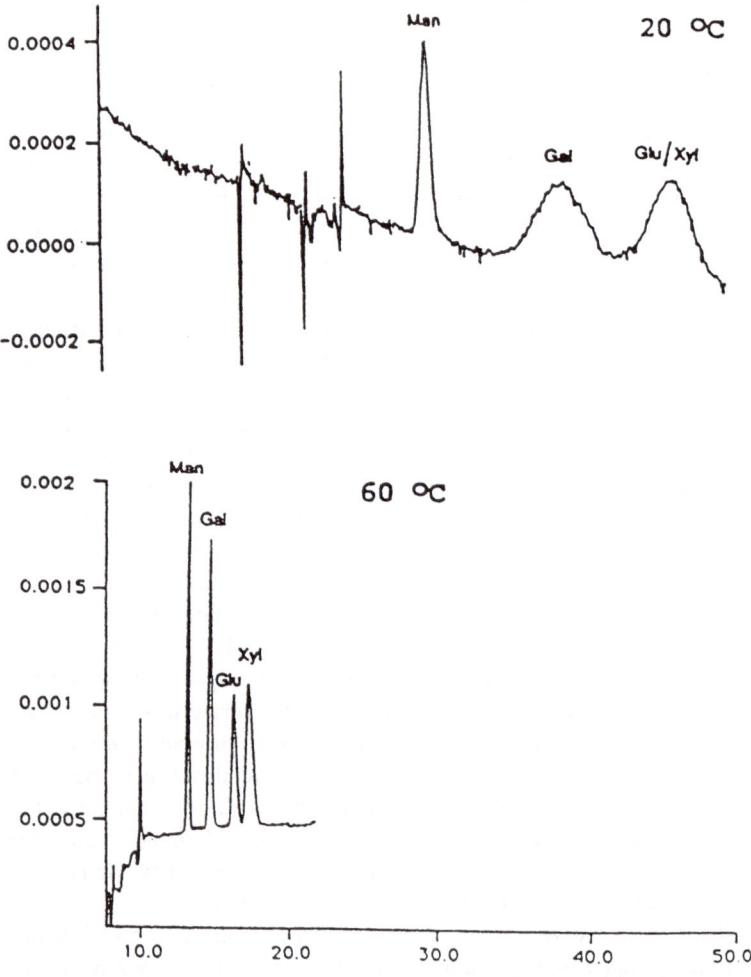

Figure 4.2 Effect of temperature on CE of non-derivatized monosaccharides. Separation conditions:
BGE 50 mM borate, pH 9.3; capillary, 94/87 cm x 75 µm ID; voltage, 20 kV; temperature,
20–60°C; UV detection, 195 nm. Sample: mannose (Man), galactose (Gal), glucose (Glu)
and xylose (Xyl) (from [7], with permission).

In contrast to that a decrease in resolution and number of theoretical plates with increasing temperature was observed for non-reducing sugars, such as saccharose and oligosaccharides of the raffinose family (α-(1,6)-galactosides linked to the glucose moiety of saccharose) [8]. This can be attributed to an increase in longitudinal diffusion at higher temperatures. Because the non-reducing oligosaccharides can not build the open chain form the negative effect of increased diffusion is not offset by an enhanced ring opening and at the same time complexation of the carbohydrates. However, both resolution and efficiency could be considerably improved for the separation of raffinose, stachyose and verbascose by increasing the concentration of sodium tetraborate in the separation buffer from 20–100 mM.

Borate complexation allows the separation of a series of homologous oligosaccharides such as α-(1,6)-linked isomalto-, α-(1,3)-linked laminara- and β-(1,4)-linked cello-oligoglycans by complexation of the outermost glucose residue [9]. Since all oligosaccharide fragments carry the same charge and their electrophoretic mobility depends on their charge-to-mass ratios, size separations are possible, with the mono- and disaccharides migrating in front of the larger oligosaccharide fragments.

4.1.1.3 CZE of glycopeptides and glycoproteins

UV detection of glycopeptides and glycoproteins is straightforward since the peptide bond allows an UV detection at 200 nm. CE separation of the major ovalbumin glycoforms could be achieved in 100 mM borate buffer, pH 8.5. Resolution was significantly enhanced through doubling the separation length, thereby effectively increasing $\Delta\mu_{ep}$ and by the addition of 1 mM putrescine to the separation buffer. Positively charged buffer additives influence the ζ-potential by binding to the surface silanol groups and therefore reducing the EOF (see equation 2.14). Enzymatic dephosphorylation of ovalbumin shifted all peaks to lower migration times indicating that all major ovalbumin isoforms are phosphorylated to the same degree. For glycopeptides the electrophoretic mobility relies on both the charged peptide moiety and the glyco part. Depending on the electrophoretic conditions/buffer choice, the migration is dictated either by the isoelectric point of the glycopeptide or by the complexation of its glycan chains with borate. The glycoforms of recombinant erythropoietin have been resolved in a triscine buffer, pH 6.2, in the presence of urea and 1,4-diaminobutane (see Figure 5.11). At low EOF, glycoforms are charge differentiated, with a migration order according to the degree of sialysation.

CZE with UV detection at 190–200 nm has also been used to investigate the carbohydrate mediated microheterogeneity of glycoproteins, for example recombinant tissue plasminogen activator (r-tPA) [10] or α1-acid-glycoprotein (AGP) [11]. High resolution separation of the hybrid- and complex type oligosaccharides of the r-tPA according to their sialic acid contents could be achieved in a phosphate or triscine buffer containing 2.5 mM putrescine as a cationic additive, causing a significant decrease in EOF (see Figure 5.20). By subjecting the enzymatically or chemically released oligosaccharides to ethanol precipitation or gel filtration prior to analysis, interference by peptides during UV detection could be minimized.

4.1.2 Electrophoresis of native carbohydrates at extreme alkaline pH

Although saccharides can be considered neutral, they can be ionized at extreme alkaline pH, forming negatively charged alcoholates. The ionization constants are very low and in the range of 10^{-12}–10^{-14}. Table 4.1 lists the pK_a values of a number of carbohydrates in water.

Electrophoresis of carbohydrates at extreme alkaline pH can be combined with three different detection methods: indirect UV detection, amperometric detection and refractive index detection.

4.1.2.1 Electrophoresis of native carbohydrates at extreme alkaline pH with indirect detection

Indirect UV detection using sorbic acid and a wavelength of 256 nm has been applied to the analysis of a number of mono-, di- and trisaccharides. All monosaccharides, found in the glycan-moiety of glycoproteins, have been separated using the same separation system (Figure 4.3)[13]. The minimum concentration to be detected with indirect UV using sorbic acid is 0.2 mM.

Using neutral monosaccharides as test compounds, Vorndran et al. [14] demonstrated that by increasing the pH of the BGE from 11.9–12.3 the resolution of the monosaccharides could be improved. Judging from the pK_a-values of the saccharides, a higher pH would result in an even greater selectivity of the separation system since a greater portion of the saccharide would be charged and therefore the separation could occur under plain CZE conditions. However, at pH greater than 12.4, the UV baseline became very noisy, effectively

Table 4.1 Ionization constants of selected carbohydrates [12]

Carbohydrate	pK_a at 25°C	Mobility in 10^{-5} [cm²/Vs]
Lactose	11.98	4.175
Maltose	11.94	4.813
D-Fructose	12.03	
D-Mannose	12.08	7.462
D-Lyxose	12.11	6.440
D-Ribose	12.21	7.419
D-Xylose	12.29	6.268
D-Glucose	12.35	5.135
D-Galactose	12.35	4.358
D-Arabinose	12.43	5.121
2-Deoxyglucose	12.52	
2-Desoxyribose	12.67	2.798
Raffinose	12.74 (18°C)	1.312
D-Mannitol	13.50 (18°C)	
D-Glucitol	13.57	
Glycerol	14.40	

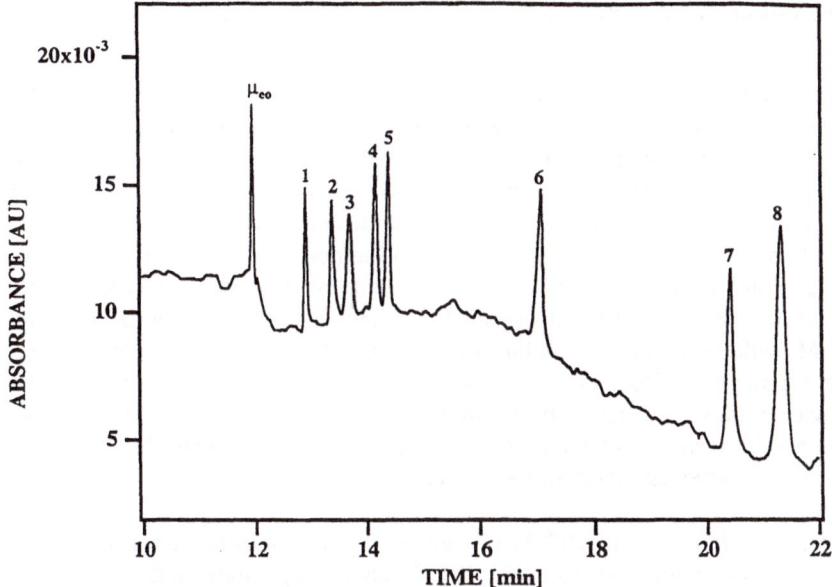

Figure 4.3 Capillary zone electrophoresis of non-derivatized carbohydrates at high pH with indirect UV detection. Separation conditions: BGE, 6 mM sorbate, pH 12.2; capillary, 90/83 cm x 50 μm ID; voltage, 230 V cm^{-1}; temperature 15°C; 1 s injection; indirect UV detection, 256 nm. Carbohydrates: 0.97–1.55 mM; 1 = fucose, 2 = galactose, 3 =glucose, 4 = N-acetylgalactosamine, 5 = N-acetylglucosamine, 6 = N-acetylneuraminic acid, 7 = galacturonic acid, 8 = glucuronic acid (from [13], with permission).

limiting analysis to pH 12.2–12.3. A variation of the temperature showed that a sub-ambient temperature of 15°C gave the best results.

Improvements in both the concentration detection limits and the selectivity could possibly be achieved by using an improved indirect absorber. The requirements for the ideal absorber, high molar extinction coefficient, an effective electrophoretic mobility close to the solutes, a single charge for a high transfer number and stability at pH 12–14, have long been know, however an indirect UV system superior to sorbic acid has not been described.

Lower detection limits could be achieved with carbohydrates carrying intrinsic negative charges. Compared to neutral compounds, these solutes displace more chromophores in the BGE. In addition, separation conditions are not confined to extreme pH ranges. In a 6 mM sorbate BGE, pH 5, D-galactonic and D-gluconic acid were baseline resolved with a detection limit as low as 18 femtomoles, corresponding to approximately 2 x 10^{-6} M. Even lower detection limits (5 femtomoles) could be obtained for a synthetic heparin fragment containing multiple carboxylic acids and sulfate groups, separated in a 5 mM sulfosalicylic acid at pH 3.

Indirect fluorescence detection using coumarin 343 as a fluorophore does not show any improvement over indirect UV detection in term of sensitivity, mainly because of laser noise. The separation is further rendered difficult because the dye degrades above pH 11.5. However, separation of monosaccharides as well as of some high molecular weight polysaccharides, such as dextran or amylose, could be demonstrated [15].

4.1.2.2 Electrophoresis of native carbohydrates at extreme alkaline pH with amperometric detection

The success of the HPAEC/PAD system for carbohydrate analysis clearly demonstrated that amperometric detection is a highly sensitive and selective detection method for non-derivatized sugars. Not surprisingly, amperometric detection in CE was developed as an alternative to insensitive photometric methods. Since amperometry as well as conductivity is pathlength independent, very narrow capillaries can be used. Zare and his group demonstrated a separation of a mixture of 15 carbohydrates at pH 13 in under 45 minutes with efficiencies of up to 200'000 theoretical plates [16]. Detection was accomplished by a cylindrical copper electrode with amperometric detection at constant potential (Figure 4.4). Concentration detection limits of 5×10^{-6} M could be achieved with linear calibration plots over three orders of magnitude. The high pH improved the selectivity of the separation system and therefore the resolution of the saccharides. Addition of borate to the BGE, to further improve the selectivity, decreased the anodic response of the carbohydrates at the copper electrode. It was assumed that the borate complexes hindered the availability of the oxidation sites present in the carbohydrates.

A slightly lower detection limit of 10^{-6} M (for glucose) could be achieved using pulsed amperometric detection at a gold wire electrode [17]. By alternating anodic and cathodic polarization to clean and reactivate the electrode surface, the problem of electrode fouling could be overcome. The uniform and reproducible electrode activity allowed the sensitive detection of glucose in human blood and of biologically important carbohydrates, such as glucosamin, glucosaminic acid, glucosamine 6-sulfate and glucosamine 6-phosphate. However, with a pH 12 BGE, containing 100 mM sodium hydroxide and 8 mM sodium carbonate, the three neutral carbohydrates saccharose, glucose and fructose, which could be resolved in other studies [14, 18] , could not be separated.

Similar results were obtained with a 10 µm disk electrode for pulsed amperometric detection. In a 10 µm capillary with a 0.1 M NaOH BGE, eight monosaccharides could be analyzed at the 10^{-6} M level [19]. Initial separation efficiencies of 100'000–200'000 decreased after a few days of operation. A selectivity increase was achieved by incorporating a borate buffer system and a post capillary reaction system for amperometric detection.

The wall-jet electrochemical detector, allowing the use of normal size working electrodes (diameter > 100 µm) represents a third approach to amperometric detection in capillaries [20]. A disk-shaped electrode consisting of a copper wire with only its tip cross section exposed was positioned immediately in front of the much smaller capillary with an opening of 25 µm. A concentration detection limit of 10^{-6} M was accomplished for mono- and disaccharides with the wall-jet design. It is worth mentioning, that neither significant post capillary zone broadening nor loss in separation efficiency was observed, although the fluid stream exciting the capillary flows radial across the electrode. The major advantages of this approach are the use of larger, more rugged electrodes, ease of use and better reproducibility of the electrode/capillary alignment.

As discussed in the HPLC literature, amperometric detection suffers from three general drawbacks independent of the detector design. The amperometric detector is non-discriminatory and positive responses originating from amino acids, peptides and organic acids can confuse peak assignment. More important, detector response is not uniform within a class of

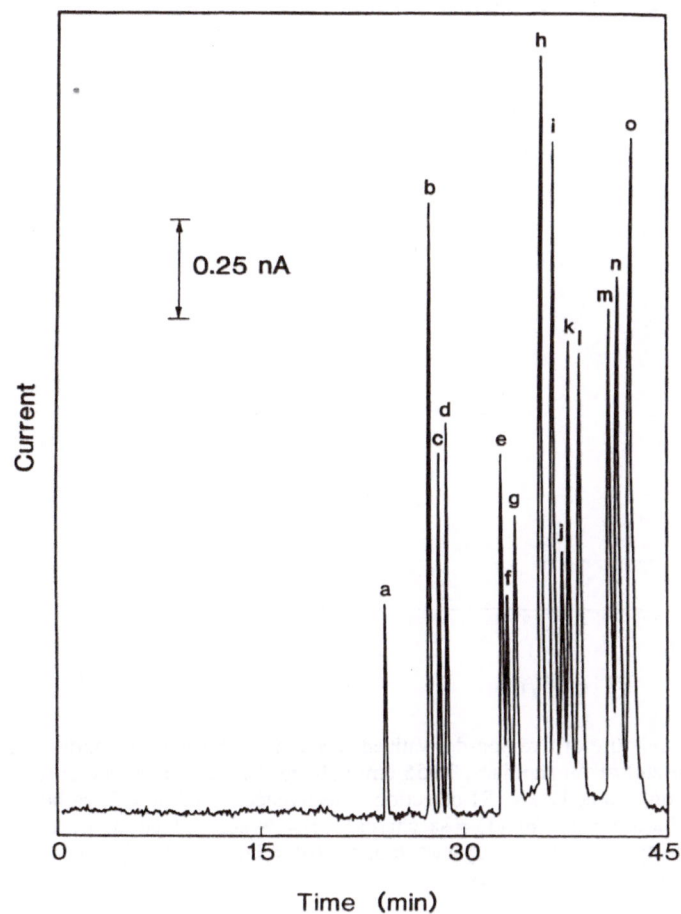

Figure 4.4 Capillary zone electrophoresis with amperometric detection of a mixture containing 15 different carbohydrates (80–150 μM). Separation conditions: BGE, 100 mM NaOH; capillary, 73 cm x 50 μm ID, voltage, 11 kV, injection, 10 s by gravity (10 cm height); peak assignment: (a) = trehalose, (b) = stachyose, (c) = raffinose, (d) = saccharose, (e) = lactose, (f) = lactulose, (g) = cellobiose, (h) = galactose, (i) = glucose, (j) = rhamnose, (k) = mannose, (l) = fructose, (m) = xylose, (n) = talose, (o) = ribose (from [16], with permission).

compounds, requiring standard curves for each solute before quantitation. Thirdly the high pH required for the proper working of the electrodes limits the choice of electrophoretic separation conditions. There is the concern, that strong alkaline conditions can induce epimerization and degradation. This is well known for reducing carbohydrates, especially if the reducing terminus is a 2-acetamido-2-deoxy sugar (N-acetylhexosamines). With 3-substituted N-acetylhexosamines at the reducing terminus, β-elimination poses a more serious problem under alkaline conditions.

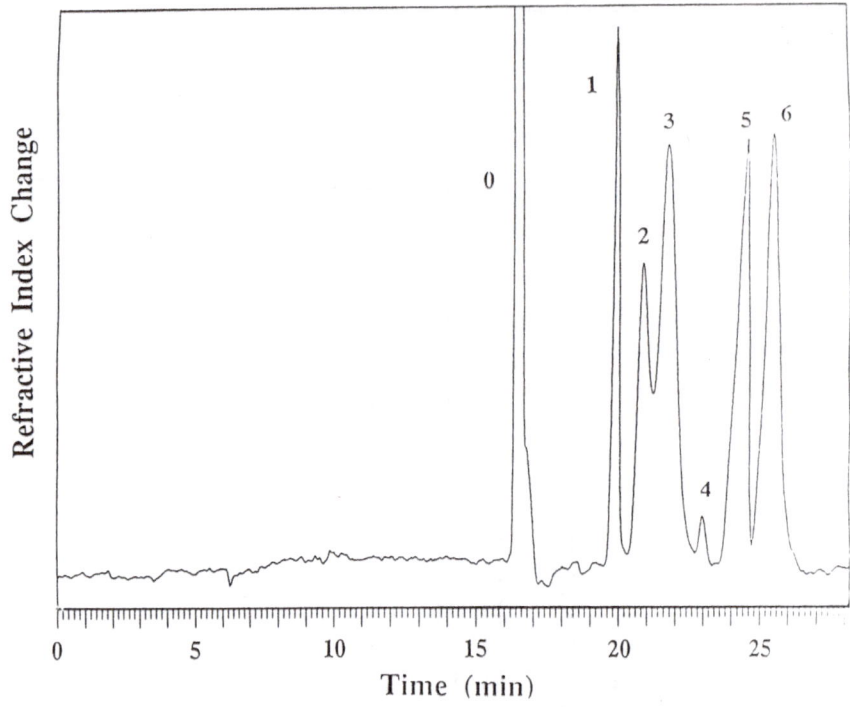

Figure 4.5 Electropherogram of a mixture of five non-derivatized saccharides. Separation conditions: BGE, 100 mM tetraborate, pH 9; capillary, 70/55 cm; voltage, 14 kV; thermocooler temperature, 27°C; injection, 7 s at 12 kV, RI-detection, interference fringe, n = 2; carbohydrates (1 % each, saccharose = 0.5 %): (1) = saccharose, (2) = N-acetylglucosamine, (3) = cellobiose, (4) = impurity, (5) = N-acetylgalactosamine, (6) = lactose. Peak (0) resulted from the BGE (from [21], with permission).

4.1.2.3 Electrophoresis of native carbohydrates at extreme alkaline pH with refractive index detection

RI detectors based on interferometry allow the detection of carbohydrates at a 10^{-4} M level after separation in an alkaline borate buffer (Figure 4.5). Since the overall sensitivity of the RI-detector depends mainly on the diameter of the capillary, higher sensitivity with a 100 μm- instead of a 50 μm capillary should be obtained. Unfortunately, the extremely high currents and the Joule heat generated inside the 100 μm-capillary at pH 13, inhibit their use. Therefore, this approach is only suitable to 10–50 μm capillaries [21–23].

4.1.3 Concluding remarks

If glycopeptides, glycoproteins or charged carbohydrates are to be separated, direct UV detection is a versatile and flexible method. The high selectivity of in-column borate complexation to attach a negative charge on the otherwise neutral polyols is also a very suitable way to perform electrophoretic analysis with carbohydrates. It is most effective at moderate alkaline pH. However, sensitivity for the direct analysis of borate complexes in the deep UV (185–195 nm) is only in the micromolar range for pure carbohydrates. For carbohydrate analysis in the food industry with less severe sensitivity requirements this method can be simple and effective.

If higher concentration sensitivities are necessary, indirect UV or amperometric detection are better suited. These detection methods are best combined with high pH electrophoresis since the neutral carbohydrates are negatively charged under these conditions. However, this approach exhibits sufficient selectivity only for mono- and some disaccharides. Indirect UV using sorbic acid is fine for medium sensitivity requirements. Amperometric detection, while slightly better than indirect UV, suffers from the fact, that it is not commercially available. Refractive index detection has only advantages when extremely small sample amounts with medium sensitivity requirements are available. Again, a lack of commercial instrumentation is limiting its use.

If both small sample amounts and high sensitivity requirements dictate the choice of the analytical system for carbohydrate analysis, a derivatization approach with a charged label might be the best alternative.

4.2 Separation and detection of derivatized carbohydrates

4.2.1 Labeling of carbohydrates

The most suitable method to convert a solute with no, or only a weak, detector response into a derivative with enhanced detector properties is pre- or post-column derivatization. The chemical characteristics of the label should be compatible with the separation mechanism and the detection scheme used. For example, the hydrophobic interaction in reversed phase chromatography favors a hydrophobic label with a high molar extinction coefficient. Since electrophoretic separations rely on charge differences, an ideal CE label should introduce both the charge to propel the solute through the capillary as well as favorable optical properties to allow sensitive UV and/or LIF detection.

Pre-column derivatization is often preferred in both LC and CE because in this case derivatization is independent of the mobile phase or the BGE, respectively and the reaction kinetics are not limiting factors. Additionally pre-column derivatization may improve the selectivity and the resolution of the overall method by changing the chemical or structural characteristics of the analyte.

Four different chemistries have been described in the literature to label carbohydrates prior to CE analysis:

(1)   Reductive amination [24],

(2)   Condensation with 1-phenyl-3-methyl-5-pyrazolone (PMP) [9, 25]

(3)   Condensation of carboxylated (acidic) saccharides with amines [26].

(4)   Esterfication of amino alditols [27]

Depending on the label characteristics, derivatization chemistry (3) and (4) can be applied for both UV and fluorescence detection.

4.2.1.1 Reductive animation

The most frequently used method for the pre-column derivatization of carbohydrates is a reductive amination. Reducing carbohydrates exist in solution in either one of two forms: an annular form or in an open chain aldehyde form. The carbonyl group can react with the amino group of a label forming a Schiff base. However, this is an equilibrium reaction with the thermodynamics favoring in most cases the educts. Therefore, in a second step, the Schiff base is reduced with sodium cyanoborohydride to form a stable secondary amine (Figure 4.6) [28, 29]. The formation of the Schiff base is the rate-limiting step, whereas the reduction proceeds rapidly. To shift the initial equilibrium towards the product and to prevent a further reaction of the final secondary amine with the carbonyl compounds, an at least five fold excess of amine is used.

Reductive amination has been used with a number of UV absorbant or fluorescent amines. Table 4.2 displays the chemical structure of the most commonly used labeling reagents for reductive amination of carbohydrates.

The optimized reaction conditions are similar for all labels. Since the reductive amination is acid catalyzed, the pH of the reaction solution affects the kinetics and the degree of the derivatization. Although the reductive amination works best in a pH range of 6–8, the reaction has been shown to work successfully in media with a pH as low as 2 and as high as pH 10. Hase et al. reported yields of 75–80% for 2-aminopyridine (2-AP) derivatized carbohydrates at pH 6.2 [30]. One of the first CE separations of reductively aminated carbohydrates was shown by Honda et al. using 2-AP as the derivatization agent (Figure 4.7) [24].

Another label, 8-amino naphthalene-1,3,6-trisulfonic acid (ANTS), was originally developed for use in polyacrylamide slab gel electrophoresis [31]. It was subsequently adapted to CE systems [32, 33]. Since ANTS was the first label reagent with both good electrophoretic and spectroscopic characteristics, the derivatization reaction using the ANTS label is examined here in greater detail. But similar conditions are also used for two other successful amine labels for reducing carbohydrates, 9-aminopyrene-1,4,6-trisulfonic acid (APTS) and 2-aminoacridone (AMAC).

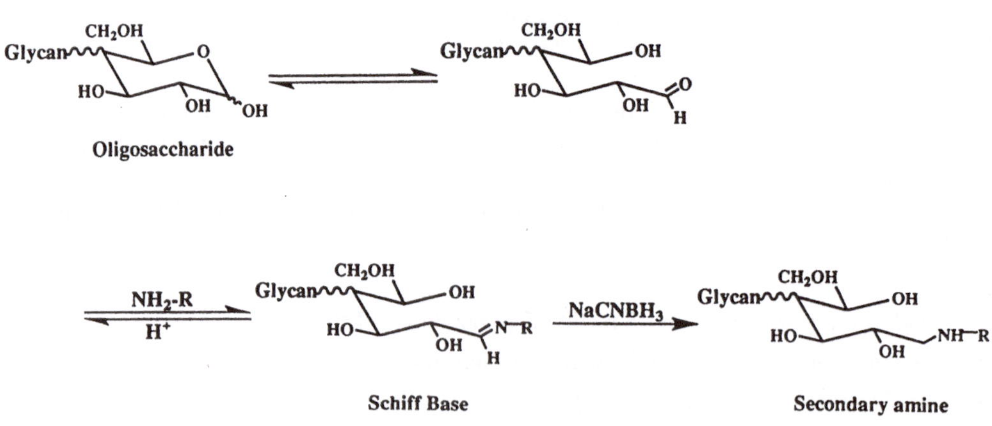

Figure 4.6 Reaction scheme reductive amination

Table 4.2 Labeling reagents for reductive amination of carbohydrates

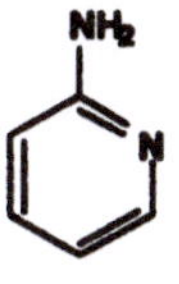

2-Aminopyridine
2-AP
λ_{max} = 240 nm

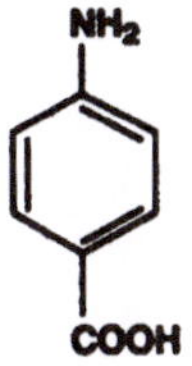

p-Aminobenzoic
acid
λ_{max} = 285 nm

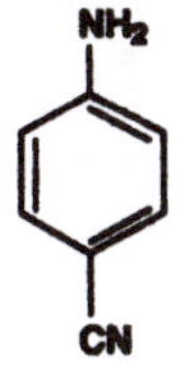

4-Aminobenzo-
nitrile
λ_{max} = 285 nm

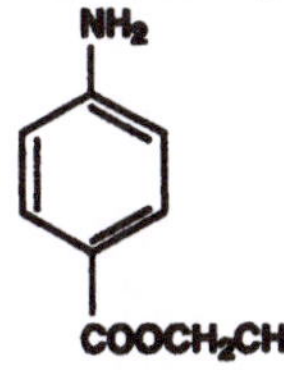

4-Aminobenzonic
acid ethyl ester
λ_{max} = 305 nm

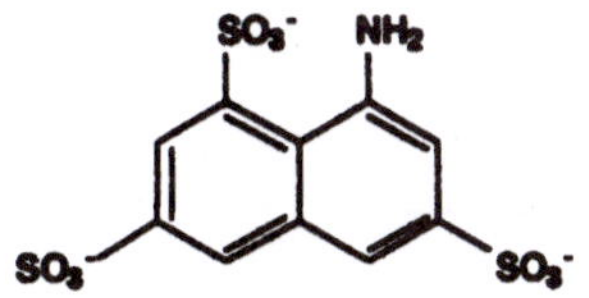

8-Aminonaphthalene-1,3,6-trisulfonic acid
ANTS
UV: λ_{max} = 223 nm
LIF: He-Cd laser: λ_{ex}: 325 nm, λ_{em}: 520 nm

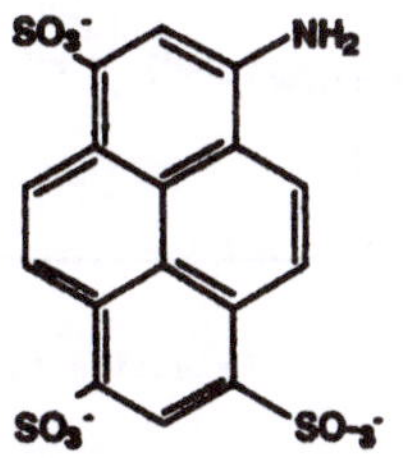

9-Aminopyrene-1,4,6-trisulfonic acid
APTS
LIF: Ar-ion laser: λ_{ex}: 488 nm, λ_{em}: 512 nm

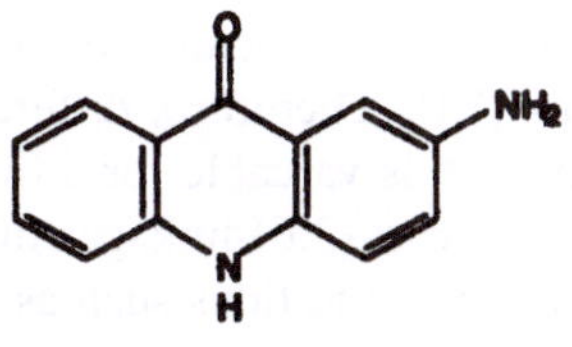

2-Aminoacridone
AMAC
LIF: Ar-ion laser: λ_{ex}: 488 nm, λ_{em}: 520 nm

In the case of ANTS, acidic derivatization conditions using 15% acetic acid and a reaction time of 15 hours at 37°C was found to result in the best yields labeling of neutral complex carbohydrates. However, the derivatization reaction proceeds also without acetic acid, but the overall yield is reduced to only 25% compared to the conditions with 15% acetic acid. Lowering the acetic acid concentration to 5% resulted in only a small decrease in the peak area to 85% [33].

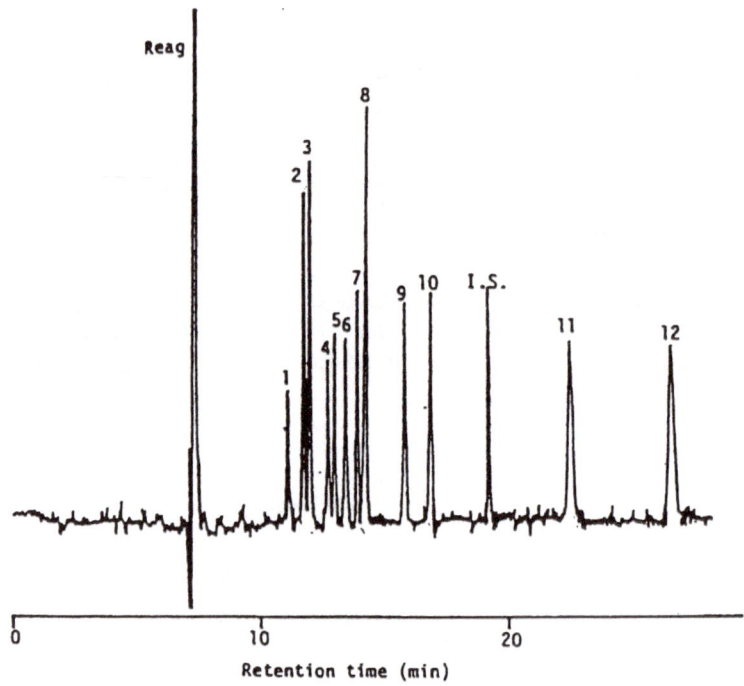

Figure 4.7 Separation of N-2-pyridylglycamines derived from various monosaccharides. Separation conditions: BGE, 200 mM borate, pH 10.5; capillary, 65 cm x 50 μm ID; voltage, 15 kV; UV detection, 240 nm. Peak assignment: Reag = reagent, 1 = N-acetyl-galactosamine, 2 = lyxose, 3 = rhamnose, 4 = xylose, 5 = ribose, 6 = N-acetyl-glucosamine, 7 = glucose, 8 = arabinose, 9 = fucose, 10 = galactose, I.S. (internal standard) = cinnamic acid, 11 = glucuronic acid, 12 = galacturonic acid (from [24], with permission).

In complex oligosaccharides the terminal sialic acids are believed to be responsible for the microheterogeneity of glycoproteins [34]. Therefore, differentiating complex carbohydrates according to their degree of sialylation is valuable for a first assessment of native microheterogeneity. Unfortunately, α-(2,3)- and α-(2,6)-linkages of the sialic acids are sensitive to acid hydrolysis. Therefore, the reaction conditions such as pH, reaction time and temperature for reductive amination were examined in respect to desialylation of complex carbohydrates. It was found, that a trisialylated complex oligosaccharide A3, (for structure see Table 5.6) did not degrade during the derivatization at 40°C for 15 hours. However, when the reaction time was speeded up to 2.5 hours at 80°C with cyanoborohydride added in a second step after 30 min, the sample decomposed resulting in 3 peaks (see Figure 4.8). Neutral complex carbohydrates stay intact during the fast derivatization conditions [35].

Another important reaction parameter is the molar ratio of labeling reagent versus sample. A 40 fold molar excess is sufficient for complete conversion of the carbohydrates into their labeled derivatives and was found optimal [33]. Below a 40-fold excess, the derivatization yield dropped. It is interesting to note that with an increase of label excess the number of side reaction products increased. Therefore the label concentration has to be carefully ad-

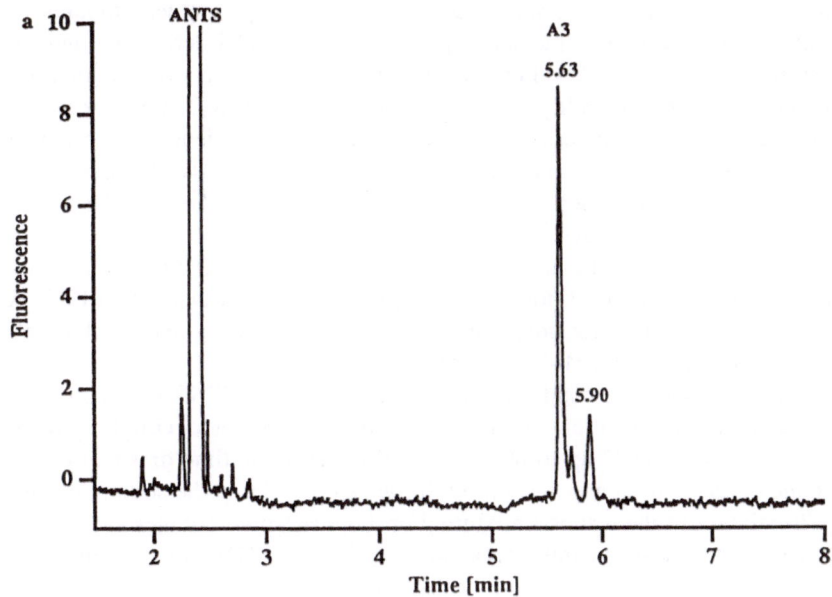

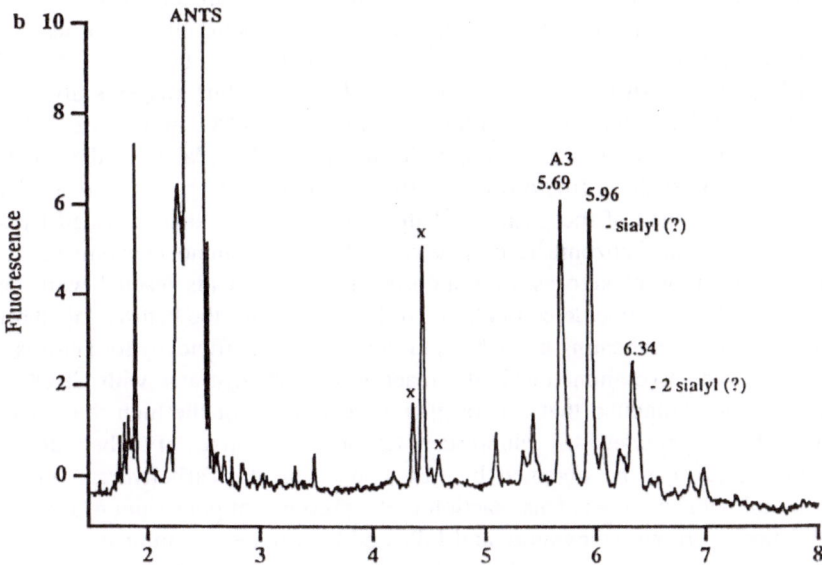

Figure 4.8 Electropherogram of a) the neutral biantennary oligosaccharide NA3 and its trisialylated analogon A3, derivatized with ANTS at 40°C for 15 h and b) the sialylated A3, derivatized with ANTS at 80°C for 2.5 h. Separation conditions: BGE, 50 mM phosphate, pH 2.5; capillary, 27/20.5 cm x 50 μm; voltage, 10 kV; temperature, 25°C; injection, 3 s, LIF detection He-Cd laser Ex 325 nm, Em: 520 nm (from [35], with permission).

justed to the sample concentration, especially when working with picomoles amounts of sample. A derivatization procedure with the above parameters for ANTS derivatization proved to work with sufficient reproducibility. For a labeling reaction of one micromole maltose under standard conditions, a reproducibility for the peak area of 4.3% R.S.D. could be achieved ($N = 8$). With glucose as a model carbohydrate a derivatization efficiency of more than 99% was found [32]. However, the reaction rate decreases for larger oligosaccharides, especially with N-acetylglucosamine residues at their reducing end, as it is the case for glycoprotein derived complex oligosaccharides

An increase in the concentration of cyanoborohydride by a factor of 20 allowed a reduction in reaction time from 15 hours to 2 hours, also helping to reduce desialylation [35]. A similar effect was observed when using boranedimethylamine as reducing agent, which allowed an almost quantitative reaction (94–96%) at 80°C.

To summarize, optimum labeling conditions could be achieved at 37°C after 10 hours. For convenience, standard derivatization times were set to 15 hours or overnight. An increase of the temperature from 37°C up to 80°C reduced the reaction time drastically to 2 hours. In this case the derivatization was carried out in two steps, adding the sodium cyanoborohydride after 30 minutes to the reaction mixture. For more sensitive samples such as the complex oligosaccharide standards, the milder reaction conditions of 37°C for 15 hours were preferred, to assure that no degradation of the sample occurred during the derivatization. Other labels such as APTS and AMAC can be attached to carbohydrates using very similar conditions with minor changes in reaction temperature and time. The major advantage of AMAC is the fast reaction time for the labeling reaction, which is performed at 90°C.

Alternatively to ANTS, a number of amino benzene derivatives were introduced as carbohydrate label, comprising 4-amino benzoic acid [36, 37], ethyl 4-amino benzoate [38] and 4-amino benzonitrile [39] (for structures see Table 4.2). These labeling reagents allowed complete derivatization of the carbohydrates within only 2 hours at 50°C, in the case of the 4-amino benzonitrile even within 15 min. An important feature of these labels is the possibility to react with ketoses, such as fructose and sorbose, which could not be labeled by 2-AP or ANTS. The importance of the position of the amino group is demonstrated by a comparison of 4- and 2-amino benzonitrile derivatives. Whereas 4-amino benzonitrile allowed successful derivatization of ketoses, only aldoses and uronic acids reacted with the 2-isomer [39]. However, the reported detection limits of 8×10^{-5} M for the ketoses fall short of detection limits for aldoses, indicating a much lower derivatization efficiency for ketoses.

A special case of a reductive amination is the reaction of carbohydrates with CBQCA (Table 4.3), a fluorogenic compound that was originally developed for the high-sensitivity analysis of amino acids and peptides [40]. Fluorescent CBQCA derivatives of carbohydrates containing no amino group can be obtained by first converting the carbohydrate into a 1-amino-1-deoxyalditol with ammonia. This reaction in the presence of potassium cyanide is completed within 1 hour at room temperature and followed by a reductive amination as described above. In this way mono- and homologous oligosaccharides as well as glycoprotein derived complex carbohydrates and glycosylaminoglycans could be derivatized [41–43]. A major drawback of this derivatization method is the closely defined molar ratio of CBQCA and potassium cyanide to carbohydrate, at which a maximum yield of CBQCA derivative can be obtained. At a one- to two-fold molar excess of CBQCA the fluorescence yield for galactosamine was highest, while it decreased to one tenth at a 0.2- and 5-fold molar excess, respectively [41]. Another disadvantage is the simultaneous labeling of other amines usually present in glycoprotein hydrolysates that may interfere with the analysis. On the other hand it is not necessary to remove the excess of CBQCA, as the non-reacted compound does not fluoresce.

4.2.1.2 Labeling by condensation with 1-phenyl-3-methyl-5-pyrazolone

The second derivatization scheme, developed by Honda and coworkers, involves the condensation between the active hydrogens of 1-phenyl-3-methyl-5-pyrazolone (PMP) with the carbonyl group of the reducing carbohydrate under slightly basic conditions, resulting in bis-PMP derivatives [9, 25, 44–46] (Table 4.3). A derivatization with 1-(p-methoxy)phenyl-3-methyl-5-pyrazolone, used at this time only with an subsequent HPLC separation, was described to be superior to the PMP for carbohydrate labeling because it is more reactive and yields a 50% higher sensitivity in UV detection at 245 nm. Since the procedure requires only slight alkaline conditions (pH 8.3) for only 30 minutes to complete the reaction, both labels are especially attractive for the derivatization of sialylated oligosaccharides because no loss of sialic acid occurs [44].

Table 4.3 Structures of non-reductive amination carbohydrate labels

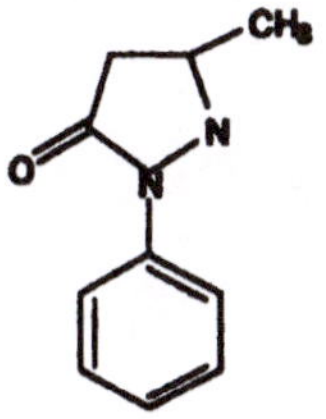

1-Phenyl-3-Methyl-5-Pyrazolon
PMP
λmax = 245 nm

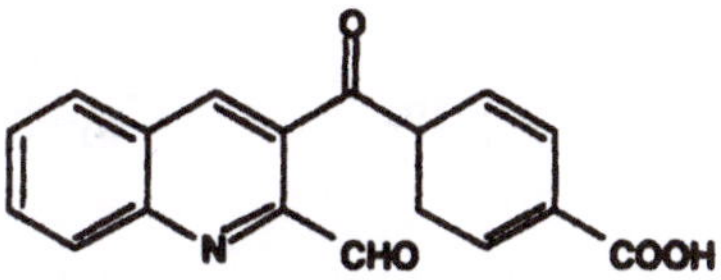

3-(4-Caroboxybenzoyl)-2quinolinecarboxy aldehyde
CBQCA
LIF: He–Cd laser: λex: 442 nm, λem: 552 nm

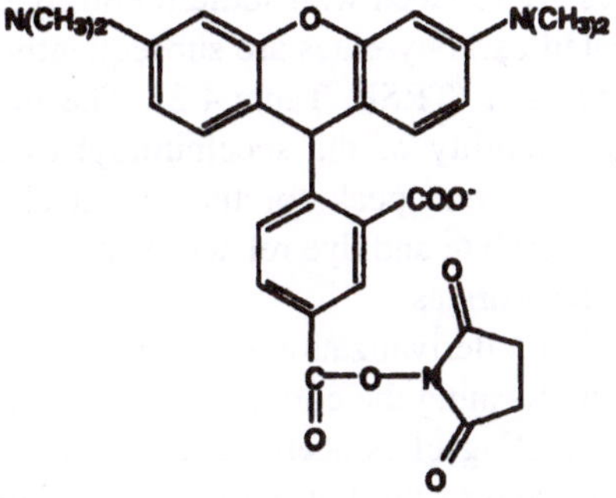

5-Carboxytetramethylrhodamine succinimidyl ester
TRSE
LIF: He–Ne laser: λex: 543 nm, λem: 580 nm

4.2.1.3 Labeling by condensation of acidic saccharides with amines

Mechref and El Rassi [26] described a new and specific derivatization for acidic monosaccharides based on the condensation between the amino group of sulfanilic acid (SA) or 7-aminonaphthalene-1,3-disulfonic acid (ANDSA) and the carboxyl group of the acidic monosaccharide to form a peptide link in the presence of a water-soluble carbodiimide (Figure 4.9). The reaction is acid catalyzed and provides not only the chromophore or fluorophore for the detection, but also replaces a weak carboxylic group of the analytes with a stronger acidic group (sulfonic acid). The derivatization of acidic monosaccharides proceeds at a fast rate. With SA, derivatization was complete within 1 hour, with ANDSA within 2.5 hours. One crucial point in the presented derivatization procedure is the amount of carbodiimide added to the reaction solution. To avoid the formation of side products the number of moles of added carbodiimide should not exceed the number of moles of reacting carboxyl groups.

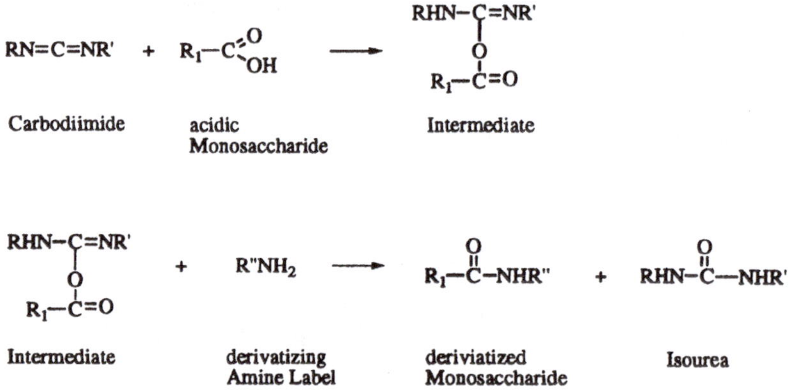

$$RN=C=NR' \;+\; R_1-C\!\!\begin{smallmatrix}\nearrow O\\ \searrow OH\end{smallmatrix} \longrightarrow \begin{array}{c}RHN-C=NR'\\ |\\ O\\ |\\ R_1-C=O\end{array}$$

Carbodiimide　　　　acidic　　　　　　　Intermediate
　　　　　　　　Monosaccharide

$$\begin{array}{c}RHN-C=NR'\\ |\\ O\\ |\\ R_1-C=O\end{array} \;+\; R''NH_2 \longrightarrow R_1-\overset{O}{\overset{\|}{C}}-NHR'' \;+\; RHN-\overset{O}{\overset{\|}{C}}-NHR'$$

Intermediate　　　derivatizing　　　deriviatized　　　Isourea
　　　　　　　Amine Label　　　Monosaccharide

Figure 4.9 Derivatization reaction for acidic monosaccharides with SA or ANDSA.

4.2.1.4 Esterfication of aminoalditols

Zhao et al. [27] introduced a two step derivatization procedure for carbohydrates. In the first step carbohydrates containing no amino group are transformed into the corresponding 1-amino-1-deoxy alditols, while hexosamines are reduced with sodium borohydride to the 2-amino-2-deoxy alditols. The resulting aminated carbohydrates are subsequently labeled with 5-carboxytetramethylrhodamine succinimidyl ester (TRSE, Table 4.3). The main drawback of this derivatization procedure is the poor stability of the succinimidyl ester in a basic buffer resulting in the generation of more than one peak for the dye itself in the electropherogram. This may cause overlapping of analyte and dye related peaks when separating more complex mixtures of TRSE labeled carbohydrates.

　　　Another approach with respect to pre-column derivatization comprises the condensation of dicarbonyl sugars with o-phenylenediamine forming the corresponding quinoxalines [47]. This reaction is especially suitable for the labeling of N-acetyl neuraminic acid, which in most other labeling schemes gives low yields due to the lower reactivity of the carboxylic group compared to the aldehyde group of aldoses.

4.2.2 Detection of labeled carbohydrates

4.2.2.1 Mass and concentration detection limits and minimal derivatization volume

An important consideration for the characterization of a derivatization procedure is not only the mass and concentration sensitivity of the labeled compounds, but also the minimum absolute amount of sample that can be labeled and detected after separation. The minimal volume, that can and needs to be handled during the sample preparation steps, is a third important parameter in method development for analytes of low concentration.

The concentration detection limit is the minimum concentration, that can be detected with a given detection principle, for example 10^{-5}–10^{-6} M with UV detection and up to 10^{-12} M with fluorescence detection. The mass detection limit is the absolute mass, subjected to an analytical separation. In the case of CE, absolute mass detection limits are rather small and in the pico- to attomole range. This is due to the miniaturization of CE with a total volume of less than 10 microliters and injection volumes in the range of a few nanoliters. However, concentration detection limits are similar to those achieved with other separation methods.

The third point of consideration is the derivatization volume. To avoid a diffusion related decrease in derivatization efficiency, the concentration of label and analyte should be as high as possible in the reaction solution. If the amount of analyte is a limiting factor, as for complex carbohydrates isolated from natural sources, the reaction volume should therefore be kept as low as possible. However, a minimum volume of 5–10 microliters is required for reproducible injection into a CE system.

In the following, mass and concentration detection limits as well as volume requirements are discussed for carbohydrate separations with CE after precolumn labeling with respect to their utility for structural and compositional analysis of glycoprotein derived complex carbohydrates.

As miniaturization of analytical instrumentation and biological insight in ever more selective systems grows, highly sensitive detection remains a pressing problem for all analytical methods discussed. Up-to-date peptide and protein sequencers need roughly 1–10 picomoles protein material to obtain unambiguous sequence information. Assuming an "average model protein" with a molecular mass of 50'000, this translates into an absolute mass of 100–1000 nanograms of sample. Since glycoproteins are usually found with a degree of glycosylation of 10–60 % by weight [48], this results in 100–600 nanograms of an oligosaccharide mixture, if all glycan chains can be completely released. Assuming further ten different oligosaccharides in the glycoprotein with an average molecular weight of 2'000, roughly 10–60 nanograms or 5–30 picomoles in absolute mass units for each oligosaccharide are available for analysis. Dissolving this amount of sample in 10 microliters, a volume that is required for reproducible injection in CE as well as for reproducible and easy to handle derivatization, this results in sample concentrations of $0.5 - 3 \times 10^{-6}$ M. Therefore, even under optimum conditions the detection limit of the analytical method needs to be in the submicromolar range.

4.2.2.2 Detection sensitivity of UV derivatized carbohydrates

UV-detection limits for most derivatizing agents are around 10^{-6} M and therefore consistent with the detection limits of other UV active compounds. There are slight differences in the wavelength and concentration detection limits as summarized in Table 4.4.

Table 4.4 UV concentration detection limit of various carbohydrate labels

Label	Detection wavelength	Concentration detection limit	Reference
2-Aminopyridine (2-AP)	240 nm	$5-8 \times 10^{-6}$ M	[13, 24]
p-Amino benzoic acid	285 nm	2×10^{-6}	[13, 36, 37]
4-Amino benzonitrile	285 nm	3×10^{-7}	[39]
4-Aminobenzoic acid ethylester	305 nm	2×10^{-6}	[13]
8-Amino naphthalene-1,3,6-trisulfonic acid (ANTS)	238 nm	5×10^{-7} M	[32, 33]
Sulfanilic acid (SA)	247 nm	2×10^{-5} M	[26]
7-Aminonaphthalene-1,3-disulfonic acid (ANDSA)	247 nm	1×10^{-5} M	[26]
1-Phenyl-3-methyl-5-pyrazolone (PMP)	245 nm	$1-5 \times 10^{-6}$ M	[9]
Quinoxalines	220 nm	1×10^{-5} M	[47]
6-aminoquinoline	270 nm	1×10^{-5} M	[49]

4.2.2.3 Detection sensitivity of fluorescent derivatized carbohydrates

Only a few applications of conventional fluorescence detection in CE have been published. With a 200 W xenon-mercury lamp as the radiation source, Mechref and El Rassi [26] detected as little as 4×10^{-7} M ANDSA derivatives of acidic carbohydrates, using an excitation wavelength of 315 nm and monitoring the fluorescence emission at 420 nm. Compared to UV detection limits, a 25 fold improvement in concentration detection limit could be achieved. Six N-acetylglucosamine-oligosaccharides with different degrees of polymerization ($n = 1-6$) were labeled with ANDSA and subsequently treated with the chitinase. Conventional fluorescence detection has also been used for the quantitation of chitooligosaccharides which are formed during the chitin breakdown by chitinase [50].

In HPLC methods fluorescent detection is in general 1000 fold more sensitive than UV detection. A low photon density in the nanoliter size detection cell hampers the modification and adaptation of commercial fluorescence detectors for their use in CE. As discussed in Chapter 2, the use of lasers for fluorescence detection in CE provides high photon density and consequently low detection limits. To obtain the best detection sensitivity the laser excitation line in fluorescence should be close to the absorption maximum of the label. ANTS has an absorption maximum of 223 nm.

Using a frequency doubled Ar-ion laser with an emission wavelength of 257 nm, detection limits as low as 10^{-9} M could be achieved for ANTS maltose [51]. Figure 4.10 demonstrates the separation of a mixture of a 50–70 nM solution of neutral complex type oligosaccharides in an acidic phosphate BGE. The peaks are very broad because the instrumental arrangement required a separation capillary of 75 cm length.

However, laser sources with excitation lines in the deep UV are for economic reasons not feasible. Since ANTS and its carbohydrate derivatives exhibit a second, although lower excitation maximum at 360 nm with an emission at 520 nm, a He-Cd laser with a 325 nm line

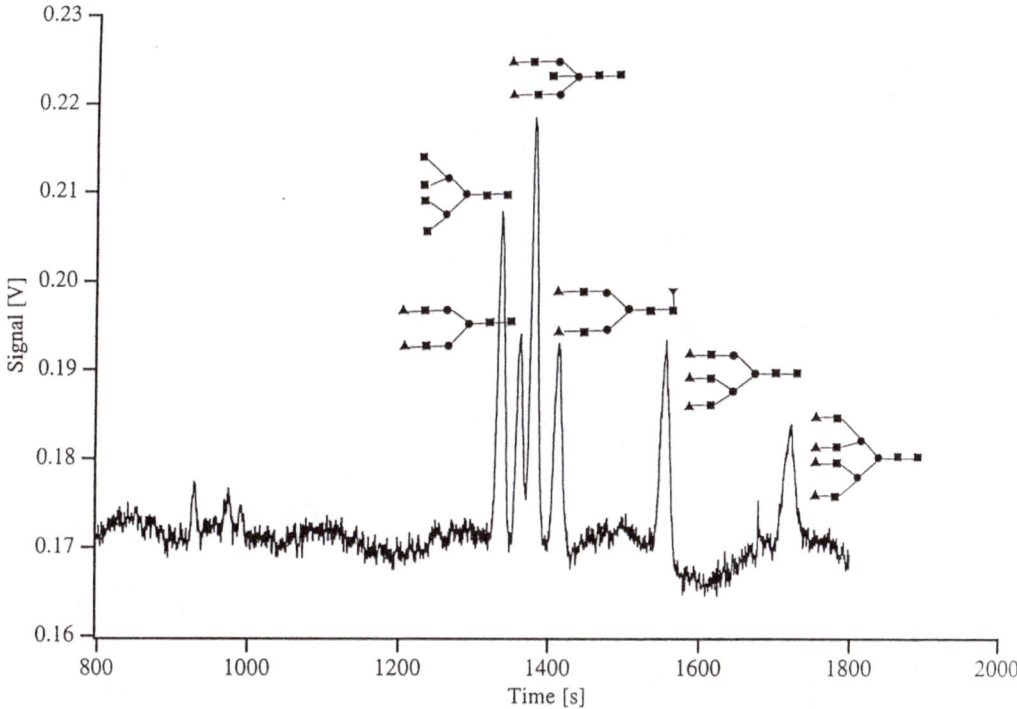

Figure 4.10 Detection of neutral complex type oligosaccharides with a double frequency Ar-ion-laser, after CE separation. Separation conditions: BGE, 50 mM sodium phosphate, pH 2.5; capillary, 70/50 cm x 50 µm ID; voltage, –20 kV; injection, 9 sec, 100 mbar; LIF detection, Ar-ion-laser, Ex. 257 nm, Em. 520 nm Samples: neutral complex oligosaccharides, 50–70 nM; ▲ = galactose, ■ = N-acetyl-glucosamine, ● = mannose, ▼ = fucose (from [51], with permission).

has been used for LIF detection with commercial instrumentation. Concentration detection limits of 5×10^{-8} M could be achieved for both small carbohydrates such as maltose [33] as well as for complex oligosaccharides with molecular weights of 2'000–3'000 [35]. Still, compared to UV detection, LIF detection of ANTS labeled carbohydrates provides only modest gains in sensitivity by one to two orders of magnitude, the reason being the poor fluorescent characteristics of ANTS with an extreme Stoke's shift of the emission band to 520 nm.

The derivatization detection limit for ANTS derivatized carbohydrates using a He-Cd laser is demonstrated in Figure 4.11. 50 picomoles of maltose, maltotetraose and maltohexaose were derivatized in a total volume of 2 microliters, containing 1 microliter of 1.5 mM ANTS and 1 microliter of 1 M NaCNBH₃. After derivatization, the carbohydrate mixture was diluted up to 20 microliter and used for injection. This corresponds to a concentration of 2.5×10^{-6} M for each sugar and an absolute amount of 20 femtomoles per peak. The mixture was separated using an acidic phosphate BGE. For comparison, the lowest detectable amount with slab gel electrophoresis is 5 picomoles per band of ANTS derivatized saccharides [31]. It is interesting to note that the LIF derivatization detection limit is almost two orders of

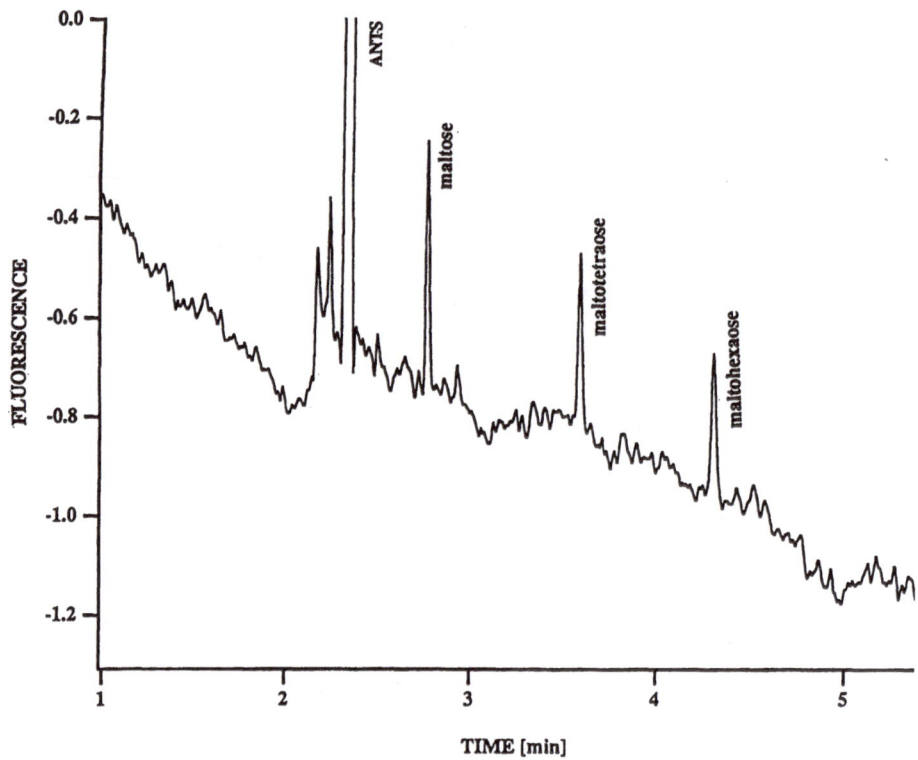

Figure 4.11 CE separation of maltose, maltotetraose and maltohexaose. Sample: 50 picomoles of
each carbohydrate. Separation conditions: BGE, 50 mM phosphate, pH 2.5; capillary,
37/30 cm) x 50 μm ID; 20 kV, 25°C; injection, 5 sec with 35 mbar; LIF detection, He-
Cd-laser, Ex. 325 nm, Em. 520 nm (from [33] with permission).

magnitude higher than the concentration detection limit. Therefore, the chemical reaction
conditions limit progress towards analyzing smaller sample amounts with higher sensitivity
and not instrumental aspects of the detection system.

Although ANTS separations clearly demonstrated the potential of fluorescently labeled
carbohydrates for glycobiology, the use of the non-standard He-Cd laser prohibited a wider
use. The cost of the He-Cd laser, lower stability as well as the need to modify the in-coming
fiber of the commercial laser to allow sufficient UV laser light for excitation proofed to be
complicated for more routine users. A labeling strategy using the commercial Ar-Ion laser is
therefore a lot more attractive.

AMAC (Table 4.2) is another carbohydrate label used in a reductive amination scheme.
AMAC can be excited either with the 442 line of a He-Cd laser or a 488 nm line of an Ar-
ion laser [52, 53]. In both cases an emission filter at 520/525 nm is used. Although no precise sensitivity data were reported typically 1 microgram in 10 microliters were derivatized.
From this information and the signal-to-noise ratios in the published literature, a concentration detection limit in the low nanomolar range can be estimated.

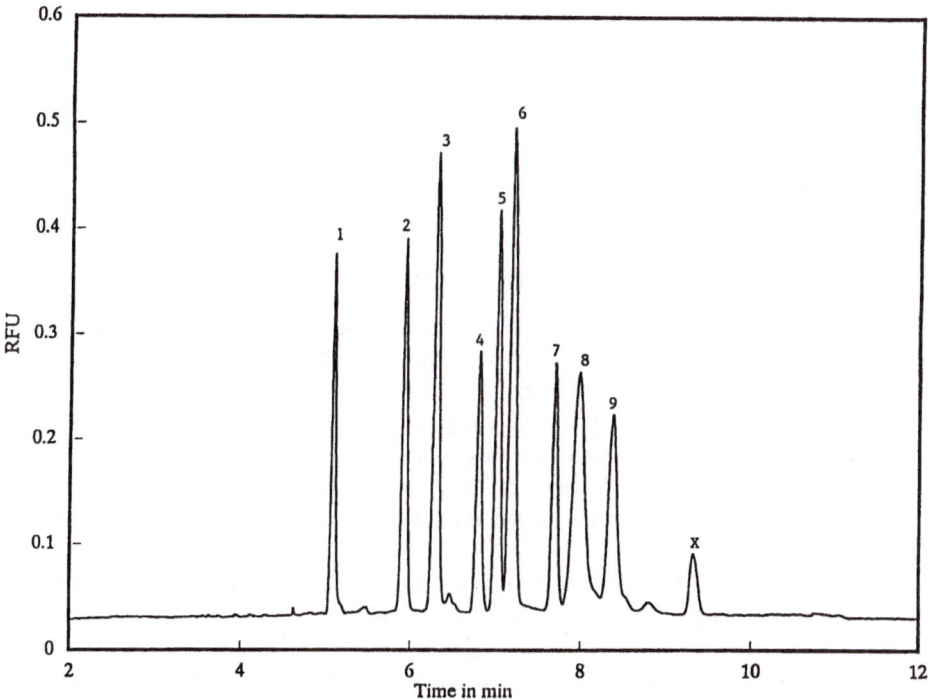

Figure 4.12 Electropherogram of nine APTS-derivatized monosaccharides at 1.0 µM each. Separation conditions: BGE, 100 mM borate, pH 10.2; untreated fused silica capillary, 27 cm x
20 µm ID; light LIF detection, Ar-Ion laser; Ex. 488 nm, Em. 520 nm, voltage, 20 kV/
19 µAs; sample injection: 20 s, 3.5 kPa pressure; outlet, cathode, Peak identification: (1)
N-acetylgalactosamine, (2) N-acetylglucosamine, (3) rhamnose, (4) mannose, (5) glucose, (6) fructose, (7) xylose, (8) fucose, and (9) galactose (X) impurity peak derived from
APTS, Outlet cathode (from [54], with permission).

APTS (Table 4.2) has been introduced by a group from Beckman Instruments to make
use of the commercial LIF detection system with an Ar-ion laser (488 nm excitation). As
little as 2 picomoles carbohydrate in 10 microliters could be detected after APTS labeling or
in terms of concentration sensitivity 2×10^{-7} M. Both monosaccharides (Figure 4.12) as well
as oligosaccharides and complex glycoprotein derived oligosaccharides were separated using
similar conditions as developed for ANTS.

The CBQCA derivatives were detectable with either an Ar-ion or a He-Cd laser-based
device. On-column fluorescence detection with the 457 nm line of an Ar-ion laser, monitoring the emission at 552 nm, resulted in detection limits of 3×10^{-9} M for CBQCA labeled
monosaccharides with a linear range over 4 orders of magnitude [40].

A He-Ne laser was used for LIF detection of the TRSE derivatives, with an excitation
wavelength of 543.5 nm and emission wavelength of 580 nm. With this set-up a concentration detection limit of 5×10^{-11} M could be achieved corresponding to a mass detection limit
of 4×10^{-22} moles (injection volume 9 picoliters) or 260 analyte molecules. This is the lowest reported concentration detection limit reached with respect to the analysis of carbohy-

drates in CE [27]. But again, not the instrumental detection limit, although extremely impressive, but the derivatization detection limit will determine how much carbohydrate can be subjected to labeling, separation and detection.

4.2.3 Separation of labeled carbohydrates

4.2.3.1 Separation in acidic electrolytes under low EOF conditions

At acidic pH, deprotonation of the silanol groups at the fused silica surface is suppressed, resulting in a very low EOF. These conditions were successfully used for the separation of a number of labeled carbohydrates. 2-AP labeled maltooligosaccharides with four to seven glucose units could be separated in a 100 mM phosphate BGE, pH 3–4.5 [55]. With a pK$_a$ of 6.7 pyridylamino-derivatives are positively charged at the acidic CE conditions employed. With the low EOF, migration order depends on the charge-to-mass ratio and is a linear function of the number of glucose residues in the homologous series, with smaller oligomers eluting first.

Since the ANTS and the ATPS label carry three sulfonic acid groups that are negatively charged over a wide pH range, these conjugates can be subjected to CE at alkaline as well as at acidic pH. Under alkaline conditions, the negative charged silanol groups on the inner capillary wall result in a strong EOF, which is usually much higher than the electrophoretic mobility of most solutes. Because optimal resolution of two solutes can be achieved by balancing the EOF against the electrophoretic migration of the solutes, resolution should be a priori improved in a separation system, exhibiting a very low or almost negligible EOF, compared to systems with a large EOF. By adding triethylamine to the pH 2.5 phosphate BGE, a masking agent for silanol groups, the EOF inside the capillary could be eliminated [32]. This way, separations with very high efficiencies compared to those achieved in alkaline BGE's were realized as depicted in Figure 4.13. Even separation times of less than 30 seconds of derivatized maltose, maltotetraose and -hexaose could be realized.

The electrophoretic mobilites of the ANTS-maltooligosaccharides in the acidic phosphate BGE were shown to be independent of the electric field strength and to increase linearly with their molecular mass as depicted in Figure 4.14, where the electrophoretic mobilites are plotted against $M_r^{-2/3}$.

Fast separations within 6–10 minutes of ANTS labeled complex oligosaccharides with excellent resolution were also achieved in a pH 2.5 BGE. The separation of the five oligomannose type carbohydrates is shown in Figure 5.25 (for structures see Table 5.6). The solutes migrate in the order of increasing molecular weight, with the smaller MAN5 eluting first and the larger MAN9 last. The peak-to-peak mass difference is 162 Daltons, equivalent to the mass of one anhydrous mannose residue. All solutes carry the same three negative charges from the ANTS label. For peak identification individual oligomannose standards were labeled, injected and their migration times compared with those of the mixture. This way the MAN5, MAN8 and MAN9 could be clearly assigned. Given the regular, oligomeric nature of the sample, the two major peaks between MAN5 and MAN8 were assigned to MAN6 and MAN7.

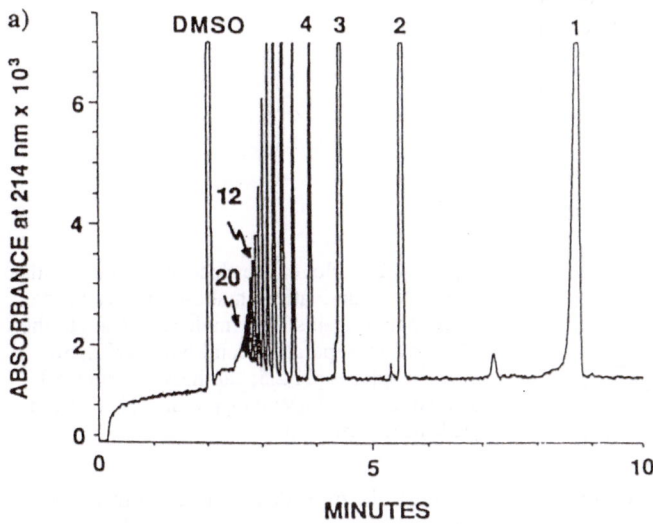

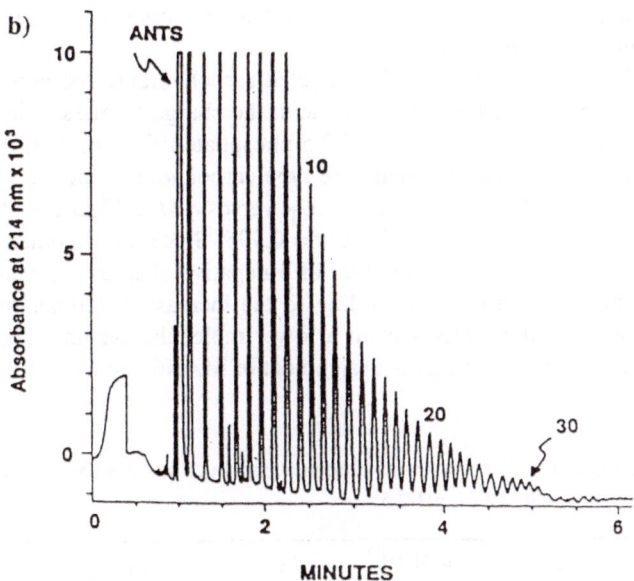

Figure 4.13 Electropherogram of ANTS derivatized malto-oligosaccharides under a) alkaline and b) acidic pH conditions. Separation conditions: BGE, a) 50 mM sodium phosphate, pH 9.0, b) 50 mM sodium phosphate, 10.8 mM triethylamine, pH 2.5; capillary, a) 37/30 cm x 50 μm ID, b) 27/20 cm x 50 μm ID; voltage, a) 17 kV, b) –22 kV; temperature 25°C; UV detection, 214 nm. Sample: a) 11 ng Dextrin 15, injected at the anodic end, b) 80 ng of Maltrin M040 starch hydrolyzate, injected at the cathodic end (from [32], with permission).

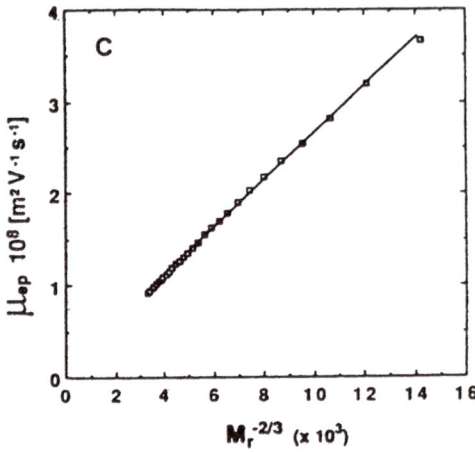

Figure 4.14 Plot of the electrophoretic mobility of ANTS-derivatized malto-oligosaccharide homologues against their molecular mass to the $-2/3$ power. Separation conditions: BGE, 50 mM phosphate, pH 2.5; capillary, 27 cm x 50 µm; voltage, -17 kV; temperature, 25°C (from [32], with permission).

According to electrophoretic theory, an ion or particle migrates in free solution in an electrical field with an electrophoretic mobility that is proportional to its electrical charge and inversely proportional to viscosity, and its hydrodynamic radius [56] (see Chapter 2). Although equation 2.3 holds true only for small symmetric spherical ions in infinite dilution, it can also be applied to larger solutes. Since the hydrodynamic radius is related to size and molecular weight, the charge-to-mass ratio (q/M), of a solute should be proportional to its electrophoretic mobility in free solution electrophoresis.

Table 4.4 lists molecular weights of the pure and ANTS-labeled carbohydrates, the number of charges, assuming -3 for ANTS and -1 for each sialic acid, the charge-to-mass ratio, the migration time and the electrophoretic mobility in the pH 2.5 phosphate BGE, for a number of complex carbohydrates. The data for the oligomannose type carbohydrates including the assumed assignments of MAN6 and MAN7 are plotted as μ_{ep} versus q/M in Figure 4.15 and yield a linear relationship with $y = 5.839 \cdot 10^{-5} + 0.074\ x$, $r = 0.995$. Since all oligomannose type carbohydrates are neutral, having the same number of 3 negative charges originating from the ANTS label, their migration order is according to the increasing number of mannose residues. A higher q/M is therefore equivalent to a lower molecular weight. The linear μ_{ep} versus q/M relationship thus confirms the peak assignment of MAN6 and MAN7.

Table 4.4 Electrophoretic mobilities (μ_{ep}) and charge-to-mass ratios (q/M) of various ANTS-labeled oligomannoses in a 50 mM phosphate BGE, pH 2.5.

	mass [g/mol]	+ ANTS [g/mol]	charge	$q/M \cdot 10^{-3}$	t_M [min]	$\mu_{ep} \cdot 10^{-5}$ [cm^2/Vs]
MAN5	1235	1600	-3	1.88	4.67	-19.3
MAN6	1398	1763	-3	1.70	4.97	-18.1
MAN7	1560	1925	-3	1.56	5.20 / 5.29	-17.3 / -17.0
MAN8	1722	2087	-3	1.44	5.58	-16.1
MAN9	1884	2249	-3	1.33	5.78	-15.6

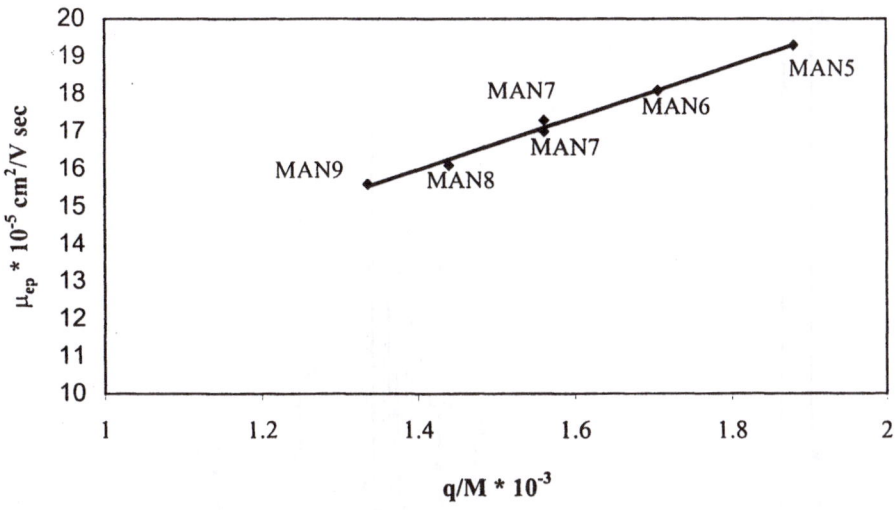

Figure 4.15 Plot of electrophoretic mobilities (μ_{ep}) and charge-to-mass ratios (q/M) of various ANTS-labeled oligomannoses. Separation conditions see Figure 5.25.

Figure 4.16 shows the separation of seven different neutral bi-, tri- and tetraantennary (NA2, NA2B, NA2F, NA3, NA4, NGA2F, NGA2FB) and two sialylated (A2, A2F) complex type oligosaccharides in the pH 2.5 phosphate BGE in less than 10 min (for structures see Table 5.6). The sialylation in A2 accounts for two additional negative charges compared to the neutral analog NA2. The additional charges result in higher charge-to-mass ratios and faster migration despite the increase in molecular weight. In accordance with the theory, the largest neutral oligosaccharide, NA4, has the longest migration time.

As discussed for the homologous series of oligomannoses, the principal separation mechanism for this set of complex type oligosaccharides is the difference in the charge-to-mass ratio. A plot of μ_{ep} versus charge-to-mass ratio (Figure 4.17) shows a close to linear relationship ($r = 0.9754$), supporting the assumption that the separation mechanism is governed by charge and mass. However, a closer inspection of the data reveals some deviations. NA2B and NA2F contain the same number of monosaccharides, with the only difference in NA2B carrying a bisecting GlcNAc instead of a Fuc residue in the core structure of NA2F. This leads to a molecular mass difference of only 57 Daltons. Nevertheless, both could be separated. Surprisingly, NA2B with the lower charge-to-mass ratio ($q/M = 1.36 \cdot 10^{-3}$), migrates in front of NA2F ($q/M = 1.39 \cdot 10^{-3}$). A similar migration inversion could be observed for the NGA2F/NGA2FB pair (Table 4.5).

On the other hand, a solute pair with a rather large difference in molecular weight, like the NGA2F/NGA2FB pair ($\Delta M = 204$), was just baseline resolved, whereas NGA2FB and NA2 ($\Delta M = 25$) migrated with significantly different migration times.

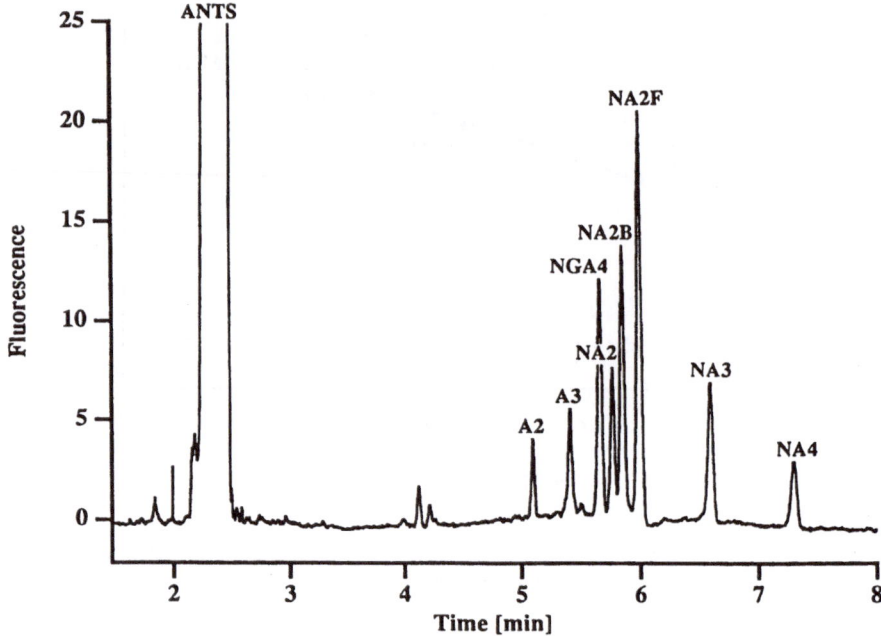

Figure. 4.16 CE separation of sialylated (A) and neutral (NA) complex type oligosaccharides. Separation conditions: BGE, 50 mM phosphate, pH 2.5; capillary, 27/20.5 cm x 50 μm; voltage, 10 kV; temperature, 25°C; He-Cd- laser LIF detection: Ex 325 nm, Em: ɔ̣0 nm (from [35], with permission).

Table 4.5 Electrophoretic mobilities (μ_{ep}) and charge-to-mass ratios (q/M) of various neutral and sialylated ANTS derivatized N-linked oligosaccharides in a 50 mM phosphate BGE, pH 2.5

	mass [g/mol]	+ ANTS [g/mol]	charge	$q/M \cdot 10^{-3}$	t_M [min]	$\mu_{ep} \cdot 10^{-5}$ [cm^2/Vs]
A2	2224	2589	−5	1.93	5.54	−17.9
A2F	2370	2735	−5	1.83	5.71	−17.4
NGA2F	1463	1828	−3	1.64	6.14	−16.2
NGA2FB	1667	2032	−3	1.48	6.30	−15.8
NA2	1642	2007	−3	1.49	6.93	−14.5
NA2B	1845	2210	−3	1.36	7.04	−14.3
NA2F	1788	2153	−3	1.39	7.28	−13.9
NA3	2007	2372	−3	1.26	8.12	−12.6
NA4	2373	2738	−3	1.10	9.21	−11.2

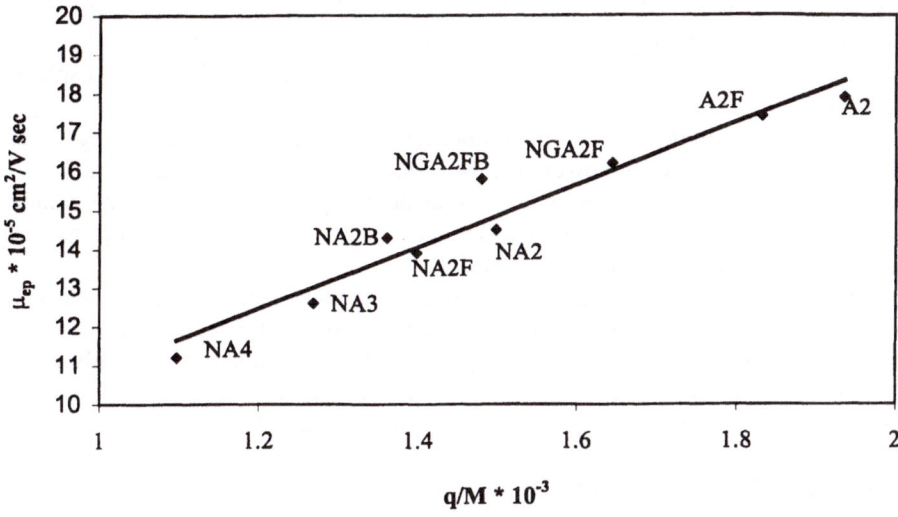

Figure 4.17 Electrophoretic mobility versus charge-to-mass ratio relationship of ANTS-labeled neutral and sialylated complex type oligosaccharides. Separation conditions as in Figure 4.16 (from [35], with permission).

These data clearly demonstrate that the charge-to-mass ratio can not be the only parameter governing the migration of complex carbohydrates. Since it is possible to separate carbohydrate isomers in acidic phosphate BGE [57] it must be assumed that their three-dimensional structure has an impact on the migration. The 3D-structure is determined by the position of the terminal sialic acids, the position of the linkage of the sugar residues, their α/β-anomerity, as well as the branching pattern of the oligosaccharides. The observation, that branched structures migrate faster than the linear ones due to their smaller effective surface area [58], supports this assumption. In a similar manner, the migration behavior of NA2B and NA2F can be explained. The additional Fuc in NA2F leads to a higher increase in the effective surface area of the NA2 molecule, due to the 1,6-linkage, than the addition of a bisecting GlcNAc. Therefore, NA2F should exhibit a higher friction coefficient than the NA2B and consequently a lower electrophoretic mobility.

A more detailed study of the resolution of ANTS-labeled monosaccharides under low EOF conditions also revealed surprising selectivity effects [59]. The seven monosaccharides found in glycoproteins - Glc, Gal, Man, Fuc, Xyl, GlcNAc and GalNAc -differ only slightly in their molecular weight. Glu, Gal and Man as well as GlcNAc and GalNAc are diastereomers, respectively. Therefore, a separation of the 7 ANTS-monosaccharides was not expected. Surprisingly, 5 peaks for a mixture with only 4 different charge-to-mass ratios were observed (see Figure 4.18a). ANTS-Glc differs in its mobility from the Gal and the Man derivative. Additionally, ANTS-GlcNAc could be separated from ANTS-GalNAc. However, ANTS-GlcNAc co-migrates with the unresolved Gal/Man pair. ANTS-Fuc with a lower molecular weight than ANTS-Glc, migrated slower than predicted and eluted after the latter one. With its lowest molecular weight, the ANTS-Xyl eluted first, as expected. Nevertheless, the separation mechanism can not be explained solely by charge-to-mass ratios. Varying the ionic strength in the phosphate BGE from 10 to 50 and 100 mM, both the migration time and resolution changed (see Figure 4.18). It has been speculated that a specific interaction between the carbohydrates and the phosphate ions modifies the apparent charge and mass of the analytes, resulting in changes in the electrophoretic mobilities. Such a phosphate-carbohydrate interaction must be specific for the individual monosaccharides.

Further investigation of these structure influences would require 3D-modeling of all monosaccharides and complex carbohydrates. Nevertheless, the mass-to-charge ratio is a rough indicator for the expected migration order in an acidic system and also for the number of sugar residues in the complex oligosaccharides. This can be very useful in the structural elucidation of glycoprotein derived oligosaccharides.

Although most low EOF separations were performed in phosphate systems, Klockow et al. [59] examined the influence of the anion in acidic buffer systems. Exchanging phosphate with citrate but using the same concentration and the same pH resulted in a slightly lower migration and broader peaks with reduced efficiency. In an alkaline system, the influence of the buffer ions on migration is more pronounced (see Chapter 4.2.2.2).

4.2.3.2 Separation under high EOF conditions

The full protonation of the silanol groups at alkaline pH results in a large ζ-potential and consequently a high EOF. Typical basic BGE's are phosphate and borate systems. Honda et al. compared with 2-AP labeled ovalbumin, human transferrin, fetal calf fetuin, human immunoglobuline G and human α-1acid glycoprotein derived oligosaccharides CE separation performance in an acidic phosphate and an alkaline borate system [60, 61]. In a pH 2.5 phosphate BGE, using a polyacrylamide coated capillary, five peaks corresponding to glycans with different degrees of polymerization from the hepta- to the undeccasaccharide could be resolved (see Figure 5.22) [60]. In contrast, the borate system was able to resolve nine peaks, an indication that the borate buffer also separates structural differences, especially in the peripheral monosaccharide residues of the oligosaccharide branches. Hence, high mannose type oligosaccharides, with the largest number of mannose residues in their outer positions, were found to complex more easily with borate. Consequently, oligosaccharides should be detected in the order complex < hybrid < high mannose type. However, the borate system failed to resolve high mannose type oligosaccharides having the same number of peripheral Man-residues such as MAN6 and MAN7 (for structures see Table 5.6). Combining acidic and borate buffer conditions, Suzuki et al. developed a 2-D map of relative electrophoretic mobilities, showing three distinct domains for high mannose-, complex- and hybrid type oligosaccharides respectively [61].

Borate complexation for both monosaccharides and complex oligosaccharides has been also been explored in order to enhance the separation of ANTS-labeled carbohydrates under basic separation conditions. Since the EOF is stronger than the electrophoretic mobilities of the individual ANTS-labeled compounds, the migration order is reversed compared to acidic conditions. A 150 mM borate buffer pH 9.5 resulted in a complete separation of all seven glycoprotein derived monosaccharides. Higher pH and borate concentrations were prohibitive due to high currents [59]. Both borate complexation and charge-to-mass-ratio dependent migration of the complex carbohydrates contribute to the migration differences. The borate ions attach a fraction of a negative charge to an ANTS conjugate. Therefore, its electrophoretic mobility increases with increasing borate complexation, resulting in a longer migration

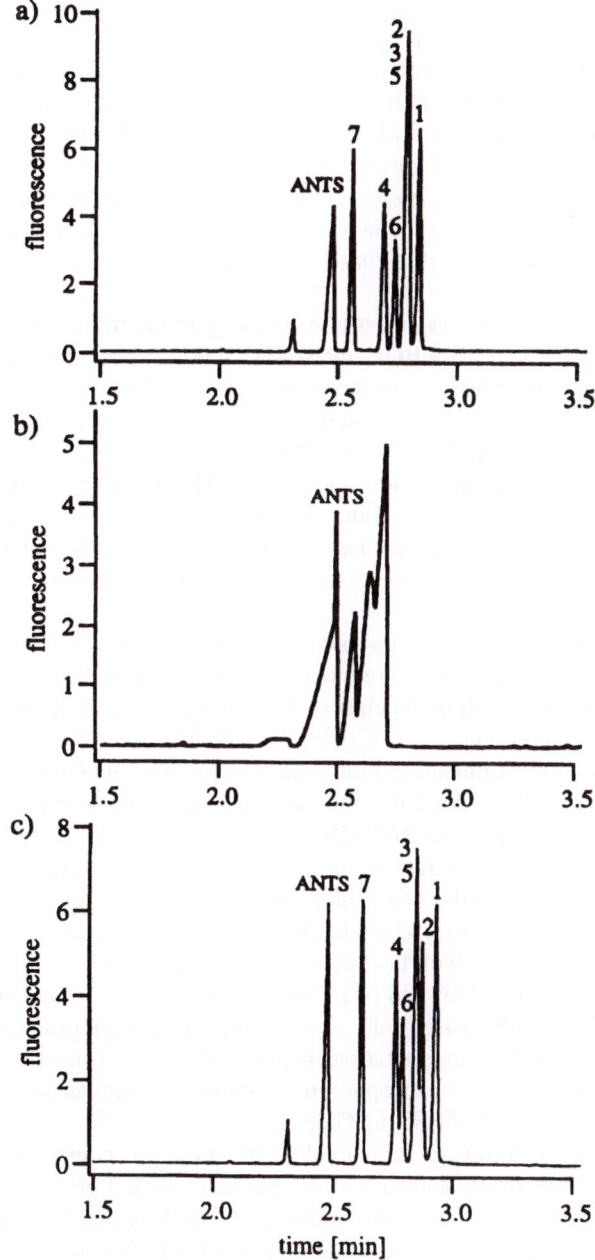

Figure 4.18 Separation of ANTS labeled monosaccharides in a phosphate BGE of different concentrations. Separation conditions: BGE, a) 50 mM, b) 10 mM, c) 100 mM phosphate pH
2.5; capillary, 27/20.5 cm x 50 µm ID; voltage, 10 kV; temperature 25°C; LIF detection,
He-Cd laser, Ex. 325 nm, Em. 520 nm. Peak assignment: 1 = GalNAc, 2 = GlcNAc, 3 =
Man, 4 = Glc, 5 = Gal, 6 = Fuc, 7 = Xyl.

time. Since the electrophoretic mobilities in borate buffers are higher than in comparable phosphate BGE's, the effective charge of the ANTS conjugates must be greater, originating from higher complexation constants. The configuration and the number of hydroxyl groups present in the monosaccharides influence the degree of borate complexation and the complex stability. For derivatized carbohydrates the open chain form can be assumed. Borate complexation in this case primarily depends on the configuration of the vicinal hydroxyl groups at C_2–C_4 and to a lesser degree at C_3–C_5, with the *threo*-diols being energetically more favorable than the *erythro*-diols. Complex stability and, hence, electrophoretic mobility increase in the following order: *erythro-erythro* < *erythro-threo* < *threo-threo* < *threo-erythro* [62].

A test mixture of 9 complex oligosaccharides was used to examine both the influence of pH and borate complexation on resolution. In a pH 9 BGE, the strong EOF leads to a fast separation in less than 4 minutes with reversed migration order, compared to acidic conditions, and a decreased resolution. Independent from the buffer composition, both borate and phosphate result in high theoretical plate numbers of 100'000–200'000 for a 20 cm effective capillary. However, the influence of the borate complexation is clearly visible by the co-migration of three compounds with 8, 10 and 11 monosaccharides residues (NGA2F, NA2F and NA3) (peaks 3/7/8 in Figure 4.18). On the other hand, two carbohydrates with the same number of monosaccharides (NGA2FB and NA2) are well separated and migrate faster than the larger NA2B with 10 sugar residues.

For complex carbohydrates it can be assumed that the outermost annular sugar residues at the non-reducing terminus of the carbohydrate chain are most accessible for complexation with borate. In the annular form, complex formation depends on the configuration of the hydroxyl groups at C_2–C_4, with *cis* diols being preferred over *trans* diols. While in the degalactosylated NGA2F and NGA2FB the outermost linkages in the sugar tree are built by GlcNAc molecules, NA2, NA2B and NA2F carry additional Gal residues at the end of both of their antennaries. Consequently, the complex stability for NGA2F and NGA2FB is diminished, due to the N-acetyl substituent at the C_2-position in the outermost GlcNAc residues, resulting in lower electrophoretic mobilities compared to the other biantennary complex carbohydrates. The lower borate complex stability of the degalactosylated biantennary oligosaccharides might be the reason why the NGA2F co-migrates with the larger galactosylated NA3.

At the additional bisecting GlcNAc in the NA2B and NGA2FB, borate complexation can take place, although not to a very high extent because of the *trans* oriented hydroxyl group at C_3 and C_4. Nevertheless, the additional borate complexation increases the overall negative charge of NA2B and NGA2B resulting in higher electrophoretic mobilities compared to its counterparts NA2/NA2F and NGA2F without the bisecting GlcNAc.

To test the buffer influence on a more "realistic" glycobiological sample in contrast to mixture of 9 known characterized carbohydrate standards, the oligosaccharide library derived from white hen egg ovalbumine, was investigated in the phosphate BGE at pH 2.5 and in the borate BGE at pH 9.5. Since the glycan part of ovalbumin consists of high mannose and hybrid-type carbohydrates, both present in a ratio of 1:1 [60, 62], this library represents a very heterogeneous mixture. Figure 4.20 shows the results of the CE separation in both buffer systems. Surprisingly, it is very difficult to correlate both traces with each other. In the phosphate system, a total of 15 peaks in a time window of 5.0–6.8 min are resolved. The borate system produced the same number of peaks within only 3.2–4.3 min. Due to the

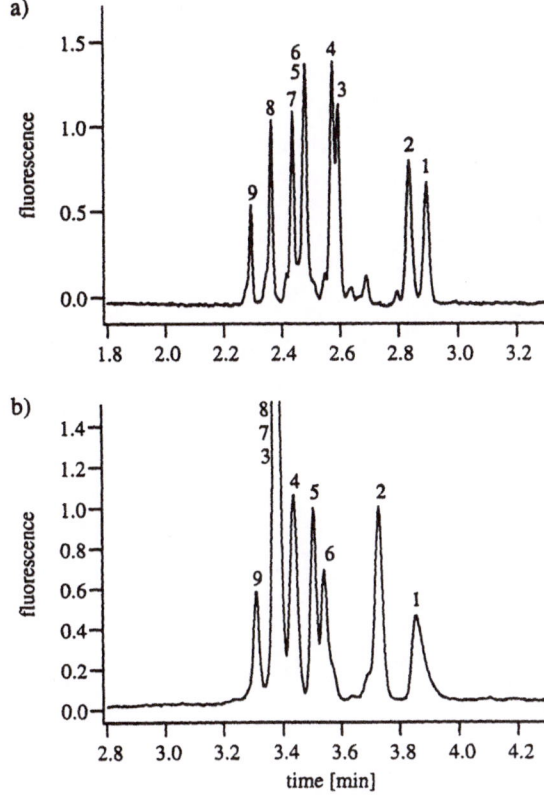

Figure 4.19 Separation of neutral and sialylated complex type oligosaccharides in alkaline buffer systems. a) 50 mM phosphate BGE, pH 9.0 and b) 150 mM borate BGE, pH 9.5, capillary, 27/20.5 cm x 50 µm ID; voltage, 10 kV; temperature 25°C; LIF detection, He-Cd laser, Ex. 325 nm, Em. 520 nm. Peak assignment: 1 = A2, 2 = A2F, 3 = NGA2F, 4 = NGA2FB, 5 = NA2, 6 = NA2B, 7 = NA2F, 8 = NA3, 9 = NA4 (from [59], with permission).

higher EOF, a possibly condensed mirror image to the phosphate BGE was anticipated. However, the influence of borate complexation greatly shifts the relative migration order.

The high mannose type carbohydrates in ovalbumin could be assigned by co-injection of the purified standards [35]. The 3 higher peaks in the early part of the phosphate electropherogram co-migrated with MAN5, MAN6 and MAN7. An unambiguous identification of MAN7 was not possible, since its concentration in the reference mixture was fairly low compared to the ovalbumin. By co-injection in the borate buffer, the same compounds could be assigned as depicted in Figure 4.20.

The lower migration rates of the high mannose-oligomers in Figure 4.20 can be attributed to their higher molecular weights and to the stronger borate complexation of the peripheral mannose, compared to the smaller ovalbumin hybrid structures with peripheral GlcNAc residues [60]. Surprisingly, the peak assigned to MAN7 in the phosphate BGE is large compa-

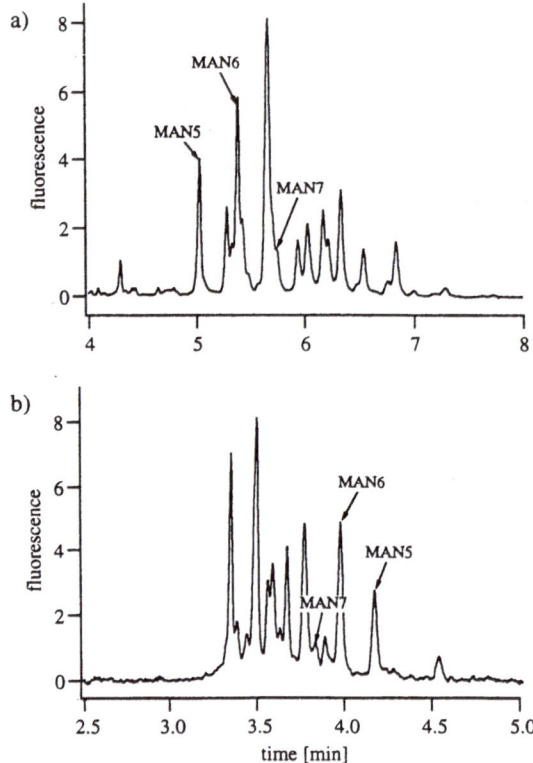

Figure 4.20 Electropherograms of ANTS derivatized ovalbumin derived glycans obtained at acidic and alkaline pH. Separation conditions: BGE, a) 50 mM phosphate BGE, pH 2.5 and b) 150 mM borate BGE, pH 9.5, capillary, 27/20.5 cm x 50 μm ID; voltage, 10 kV; temperature 25°C; LIF detection, He-Cd laser, Ex. 325 nm, Em. 520 nm (from [59], with permission).

red to the rather small MAN7 peak in the borate BGE. This is further evidence, that the preliminary peak assignment in the phosphate system is partly incorrect. Another complex carbohydrate compound, present in a considerably larger amount than MAN7, co-migrates with it, indicated by the slight shoulder in the assigned peak. Only the independent separation mechanism in the borate buffer was able to reveal this superposition.

In contrast to the findings with the standard complex oligosaccharides, where the acidic phosphate is superior to the borate BGE, the ovalbumin oligosaccharide pattern shows separations with similar resolution and the same number of peaks in both buffer systems. It can be speculated, that in borate buffers, isomeric carbohydrates are separated with a higher resolution than in an acidic phosphate buffer. On the other hand, separation mechanism in the borate buffer is based on both, borate complexation and charge-to-mass ratios. Both mechanisms are independent from each other. Therefore, in complex mixtures, different sets of solutes co-migrate in the phosphate and the borate system.

Comparison of the separation pattern obtained with ANTS with the one obtained with 2-AP labeled ovalbumin derived oligosaccharides [60] reveals, that the ANTS based system is superior in terms of resolution and speed of analysis. In an acidic phosphate and a borate BGE only 5 and 9 2-AP labeled ovalbumin oligosaccharides could be separated, respectively, instead of the 15 peaks that appeared after separation of the ANTS-labeled oligosaccharides under comparable conditions. This demonstrates that the accurate choice of the derivatization agent is of great importance in the optimization process.

Stefansson and Novotny [63, 64] demonstrated the suitability of the ANTS derivatization for separation and detection of various oligo- and polysaccharides, differing in the degree of polymerization. The separations were carried out in polyacrylamide coated capillaries using 100–200 mM Tris-borate buffers, pH 8.65. The authors stress the importance of using wall-coated capillaries to eliminate the EOF. In non-coated fused silica capillaries, the EOF is much higher and reversed in direction to the electrophoretic mobility of the labeled oligosaccharides. With increasing size, the net velocity approaches the EOF velocity asymptotically, thereby decreasing resolution with size. Moreover, peak efficiencies have been found to be considerably higher in coated capillaries [63]. The combination of coated capillaries and a charged label resulted in highly efficient and extremely fast separations of oligo-and polysaccharides.

Surprisingly, the effect of the charged fluorescent label on migration was more pronounced than expected, with only one fluorescent label attached to each oligomer, which in turn should be multiply charged through borate complexation. However, dextrans are known to complex poorly with borate, because of the trans orientated hydroxyl groups in the glucose units, the effect of sulfonic groups on the electrophoretic mobility appears to be strong. Consequently the migration times for the individual ANTS oligomers were considerably lower than those for the derivatives of the monosulfated analogue, resulting in better resolution, separation efficiencies with 1 million theoretical plates/m and shorter analysis times [63].

To summarize, ANTS-labeled monosaccharides are only fully resolved if one can take advantage of the borate complexation in alkaline buffer systems. In contrast, the low pH phosphate system yields overall the best results in terms of resolution and selectivity for complex carbohydrates. The alkaline phosphate BGE shows a comparable separation pattern if the EOF can be suppressed. At high EOF, the effective migration is too fast to allow complete resolution. The complexation with borate provides additional selectivity and changes the separation mechanism. Nevertheless, resolution is inferior compared to the phosphate system.

The CE separation systems used for amino benzoic acid derivatives labeled carbohydrates were also developed with the use of borate buffers. Either 175 or 150 mM borate, pH 10–10.5 was used under CZE conditions [38]. As expected from the previous discussion, the electrophoretic mobilities of the solutes were found to depend on the stability of their borate complexes. Complex stability is a function of the number of hydroxyl groups and their configuration at the C_3 and C_4 in aldoses and at the C_4 and C_5 in ketoses as well as on the presence of substituents. Derivatization does not only improve detection sensitivity but also the resolution in the borate based separation systems because the derivatized carbohydrates are forced into the open chain form. It was assumed that this was a result of changes in the conformation of the hydroxyl groups and consequently in the degree of complexation due to the attachment of the labeling molecule.

Similarly, mixtures of PMP-aldopentoses and -hexoses as well as PMP derivatized homologous oligoglucans of the isomalto, laminara and cello series could be well separated in

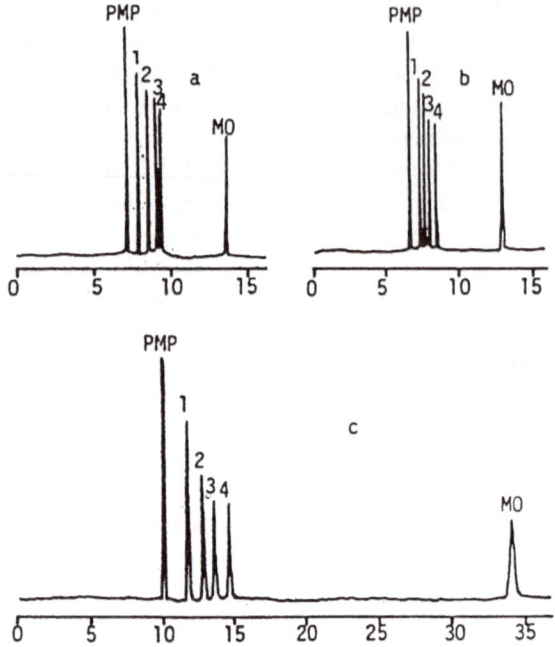

Figure 4.21   Separation of pentose-PMPs in 100 mM aqueous solutions of calcium acetate (a), barium acetate (b) and strontium acetate (c). Other Separation conditions: capillary, 49 cm x
50 µm I.D., voltage, 10 kV; UV detection, 245 nm. Peak assignment: 1 = ribose-PMP, 2
= lyxose-PMP, 3 = arabinose-PMP, 4 = xylose-PMP, MO = mesityl oxide (internal neutral marker) (from [65], with permission).

a 200 mM borate buffer, pH 9.5 [9]. Honda et al. [65] also developed an alternative separation method based on the interaction of PMP carbohydrates with bivalent metal ions. By addition of calcium ions to an acetate BGE, five PMP monosaccharides could be resolved
which co-migrated in a single peak in a sodium acetate BGE. Separation of PMP pentoses in
electrolyte solutions carrying different kinds of bivalent metal salts such as calcium-, bari-
um-, and strontium acetate resulted in varying migration times and peak resolution, as depicted in Figure 4.21. The authors attributed this to differences in ease of complexation due to
differences in electronegativity and valence angles among the metal nuclei. While these isomeric pentoses could be separated completely in the barium acetate electrolyte, separation of
N-acetylglucosamine/N-acetylgalactosamine and galactose/fucose as well as of the cellobio-
se/melobiose and gentobiose/lactose couples failed.

4.2.3.3 MECC separations

Whereas CZE separations under both acidic or alkaline conditions depend on charge differences of the solutes, MECC is more a chromatographic technique relying on differences in
hydrophobicity. Therefore, selectivity changes for the separation of 4-amino benzonitrile
derivatives from a CZE and a MECC system [39]. Generally, carbohydrates are too hydrophilic to be solubilized in ionic surfactants. But pre-column derivatization with 4-amino ben-

zonitrile rendered them hydrophobic enough to permit their separation due to the different distribution of the derivatives between the aqueous mobile phase and the micellar phase. Figure 4.22 demonstrates the MECC separation of a mixture of 16 mono- and oligosaccharides in a 25 mM Tris/phosphate BGE, pH 7.5, and 100 mM SDS. The migration times of the 4-amino benzonitrile derivatives increased in the order of ketohexoses < di-and trisaccharides < aldohexoses < deoxyaldohexoses < aldopentoses. Whereas in CZE glucose and fructose could not be baseline resolved, irrespective of the labeling molecule, complete separation could be achieved in the MECC modus. Additionally CZE failed to separate rhamnose, xylose, ribose and mannose from maltose, sorbose, lactose and arabinose, respectively, while they were well resolved by MECC.

TRSE-derivatized carbohydrates could be separated in an alkaline borate-phosphate-SDS BGE. The addition of phenyl boronic acid was found to have a great impact on the resolution of the labeled saccharide isomers, as shown for the six major hexoses found in mammalian glycoproteins.

The separation of AMAC-labeled compounds also requires a MECC system, because AMAC is a neutral label and the separation mechanism has to be based on differences in hydrophobicity. Camilleri et al. studied the resolution of AMAC-labeled complex oligosaccharides using an alkaline borate buffer system containing SDS and/or taurodesoxycholate [53, 52, 66]. The authors obtained very good selectivity, with larger fragments of the dextran ladder eluting first and the monosaccharide glucose eluting last (Figure 4.23c). The resolution decreases with the number of glucose units, but is sufficient in the size range of complex oligosaccharides with 8–11 glucose units to resolve glycoprotein derived complex oligosaccharides as demonstrated in Figure 4.23 (a and b) [53]. Especially when used in combination with HPLC and mass spectrometry, a fast and comprehensive structural elucidation of the glyco-part in a glycoprotein is feasible [52].

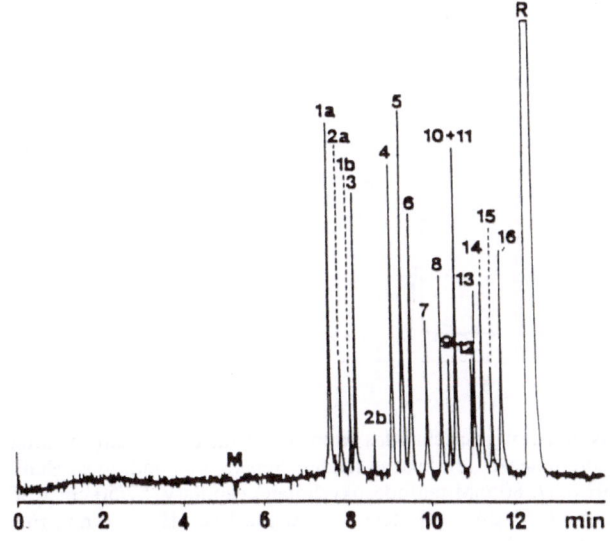

Figure 4.22
MECC of a mixture of mono- and oligosaccharides derivatized with 4 aminobenzonitrile. Separation conditions: BGE, 25 mM Tris-phosphate, pH 7.5, 100 mM SDS; capillary: 80/60 cm x 50 µm ID; voltage, 30 kV; current, 33 µAs; detection, UV 285 nm; temperature, 30°C; injection, vacuum, 1 s. Peak identification: M = methanol, 1 a/b = fructose, 2 a/b = sorbose, 3 = lactose, 4 = melibiose, 5 = cellobiose, 6 = maltotriose, 7 = maltose, 8 = mannose, 9 = glucose, 10 = glactose, 11 = ribose, 12 = rhamnose, 13 = lyxose, 14 = arabinose, 15 = fucose, 16 = xylose, R = reagent (from [39], with permission).

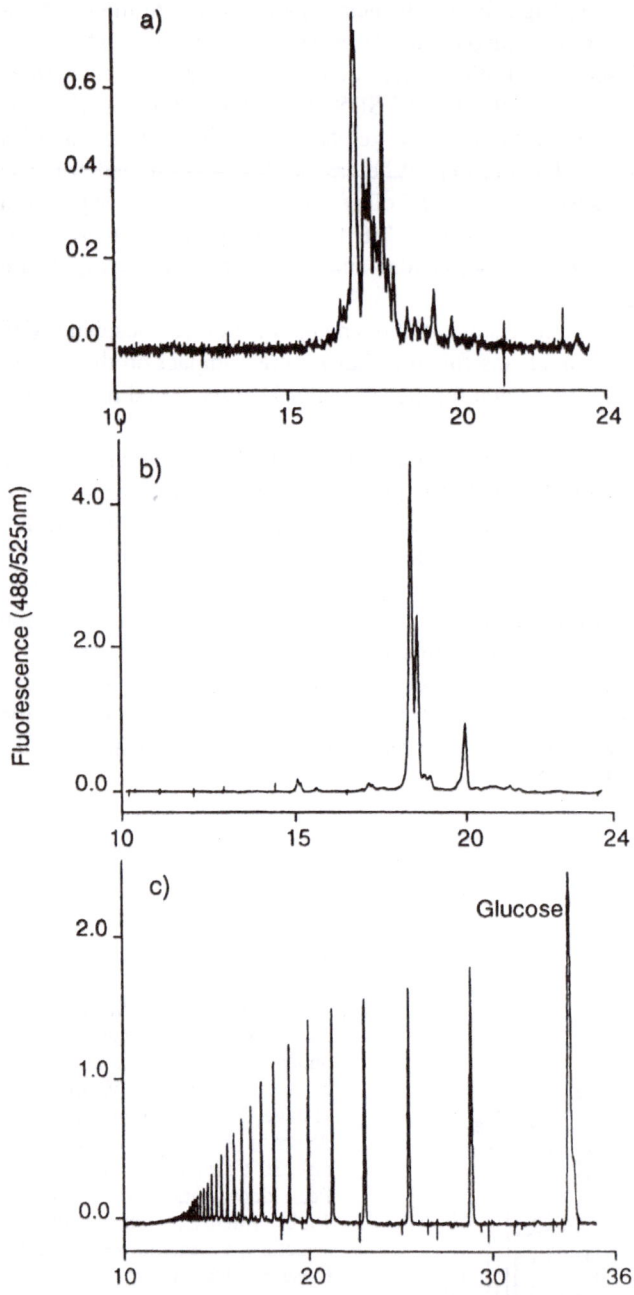

Figure 4.23 Electropherograms of derivatized oligosaccharides from fetuin (a) before and (b) after treatment with N-acetyl neuramindase; (c) dextran ladder; Separation conditions: BGE, 500 mM sodium borate, pH 8.87, 80 mM taurodesoxycholate; capillary, 57/50 cm x 50 µm ID; voltage, 20 kV; current, 50 µAs; LIF detection Ar-ion laser, Ex. 488 nm, Em. 525 nm (from [53], with permission).

4.2.3.4 Gel Separations

CBQCA labeled oligosaccharides have been separated in open tube capillaries and in capillaries filled with high concentrated (30% T and 3% C) polyacrylamide gels [42]. Figure 4.24 shows the separation of a polygalacturonic acid hydrolyzate after derivatization with CBQCA. Since in a wall-treated, gel-filled capillary no EOF occurs, electromigration of the labeled oligosaccharides depends solely on their size and charge. The separation of CBQCA labeled polydextrans with molecular weights ranging from 8'000–2'000'000 was achieved in an entangled polyacrylamide solution with applied pulsed field alteration. At constant potentials of 500 and 300 V/cm, the polysaccharides could not be size separated due to molecular stretching. By applying a potential gradient along the separation capillary which is periodically inverted at a 180° angle, the reptation behavior of the polysaccharides can be overcome, as the molecules undergo shape transition and the separation be can accomplished [43].

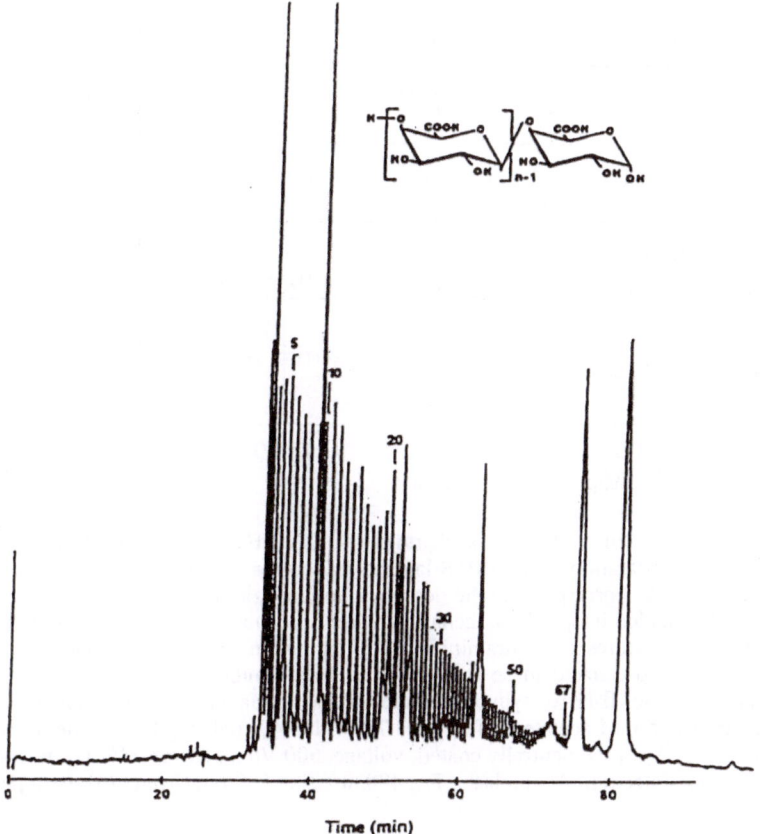

Figure 4.24 Capillary gel electrophoresis of oligomers derived from an autoclave hydrolysis of polygalacturonic acid. The numbers indicate estimated DP. Separation conditions: BGE, 0.1 M Tris/0.25 M borate/2 mM EDTA, pH 8.48; capillary, 32/23 cm x 50 µm ID; polyacrylamide gel concentration, 18% T, 3% C; voltage, 7.5 kV; electrokinetic injection, 5 kV, 25 s (from [42], with permission).

In a series of papers, Guttman et al. used gel filled capillaries to resolve mixtures of APTS-labeled carbohydrates [67–71]. Although the gel or polymeric sieving matrix used is not likely to contribute to a size separation of the oligosaccharides, it does provide a zero EOF environment. The negatively charged APTS-compounds will migrate according to their charge-to-mass ratio, similar as described in Chapter 4.2.2.1. A model for high resolution profiling based on migration times as a function of molecular size was developed [72]. Using a series of glycosidases in a predetermined matrix with a subsequent separation by CE resulted in a complete structural elucidation of the glyco-part. Figure 4.25 shows the electropherograms obtained after a set of glycosidases chopped off the glycans from the non-reducing end [73].

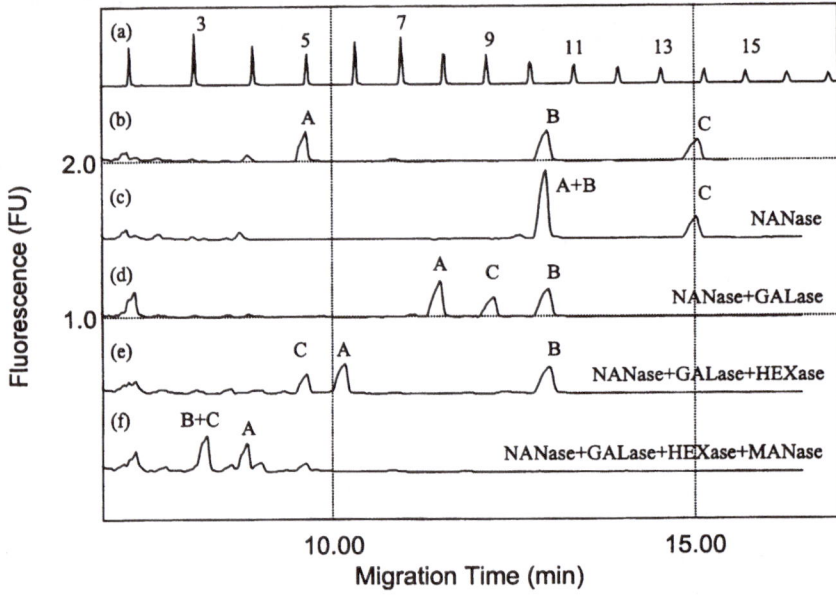

Figure 4.25 Multistructure sequencing of oligosaccharides by CE-LIF/enzyme matrix digestion assay. (a) CE-LIF separation of the APTS-labeled maltooligosaccharide ladder standard. Numbers above peaks correspond to the degree of polymerization; (b) CE-LIF separation of the APTS-labeled intact oligosaccharide mixture; (c-f) after the application of the different enzyme mixtures of Neuraminidase (c), Neuraminidase and β-Galactosidase (d), Neuraminidase, β-Galactosidase and β-N-Acetylhexosaminidase (e) and Neuraminidase, β-Galactosidase, β-N-Acetylhexosaminidase and α-Mannosidase (f). Separation conditions: BGE, 25 mM acetate buffer, pH 4.75, with 0.4% polyethylene oxide; capillary, 47/40 cm x 50 µm ID, neutrally coated; voltage, 500 V/cm, current, 19 µA; temperature, 20°C; LIF detection, Ar-ion laser, Ex: 488 nm, Em: 520 nm (from [73], with permission).

References

1. A. Pervin, A. Al-Harkim, and R.J. Linhardt, "Separation of glycosaminoglycan-derived oligosaccharides by capillary electrophoresis using reversed polarity", *Anal. Biochem.*, 221 **1994** 182-188.

2. J.B.L. Damm, G.T. Overklift, B.W.M. Vermeulen, C.F. Fluitsma, and G.W.K. van Dedem, "Separation of natural and synthetic heparin fragments by high-performance capillary electrophoresis", *J. Chromatogr.*, 608 **1992** 297-309.

3. S.A. Ampofo, H.M. Wang, and R.J. Linhardt, "Disaccharide Compositional Analysis of Heparin Using Capillary Zone Electrophoresis" *Anal. Biochem.*, 221 **1991** 182.

4. R.J. Kerns, I.R. Vlahov, and R.J. Linhardt, "Capillary electrophoresis for monitoring carbohydrates", *Carbohydr. Res.*, 267 **1995** 143-152.

5. K. Kakehi, A. Susami, A. Taga, S. Suzuki, and S. Honda, "High performance capillary electrophoresis of O-glycosidically linked sialic acid-containing oligosaccharides in glycoproteins as their alditol derivatives with low-wavelength UV monitoring", *J. Chromatogr. A*, 680 **1992** 209-215.

6. J. Boeseken, "The use of boric acid for the determination of the configuration of carbohydrates", *Adv. Carbohydr. Chem.*, 4 **1949** 189-210.

7. S. Hoffstetter-Kuhn, A. Paulus, E. Gassmann, and H.M. Widmer, "Influence of borate complexation on the electrophoretic behavior of carbohydrates in capillary electrophoresis", *Anal. Chem.*, 63 **1991** 1541-1547.

8. A.M. Arentoft, S. Michaelsen, and H. Sorensen, "Determination of oligosaccharides by capillary zone electrophoresis", *J. Chromatogr.*, 652 **1993** 517-524.

9. S. Honda, S. Suzuki, A. Nose, K. Yakamoto, and K. Kakehi, "Capillary zone electrophoresis of reducing mono- and oligosaccharides as the borate complexes of their 3-methyl-1-phenyl-2-pyrazolin-5-one derivatives", *Carbohydr. Res.*, 215 **1991** 193-198.

10. E. Watson, and F. Yao, "Capillary electrophoretic separation of recombinant granulocyte-colony-stimulating factor glycoforms", *J. Chromatogr.*, 630 **1993** 442-446.

11. M. Taverna, A. Baillet, D. Biou, M. Schluter, R. Werner, and D. Ferrier, "Analysis of carbohydrate-mediated heterogeneity and characterization of N-linked oligosaccharides of glycoproteins by high performance capillary electrophoresis", *Electrophoresis*, 13 **1992** 359-366.

12. Z. El Rassi, and W. Nashabheb, "High performance capillary electrophoresis of carbohydrates and glycocojugates", in "Carbohydrate Analysis". (ed. Z. El Rassi), Elsevier, **1995**, 267-360.

13. P.J. Oefner, A.E. Vorndran, E. Grill, C. Huber, and G.K. Bonn, "Capillary zone electrophoretic analysis for carbohydrates by direct and indirect UV detection", *Chromatographia*, 34 **1992** 308-316.

14. A.E. Vorndran, P.J. Oefner, H. Scherz, and G.K. Bonn, "Indirect UV detection of carbohydrates in capillary zone electrophoresis", *Chromatographia*, 33 **1992** 163-168.

15. M.S. Richmond, and E.S. Yeung, "Development of laser-excited indirect fluorescence detection for high molecular weight polysaccharides in capillary electrophoresis", *Anal. Biochem.*, 210 **1993** 245-248.

16. L.A. Colon, R. Dadoo, and R.N. Zare, "Determination of carbohydrates by capillary zone electrophoresis with amperometric detection at a copper microelectrode", *Anal. Chem.*, 65 **1993** 476-481.

17. T.J. O'Shea, S.M. Lunte, and W.R. LaCourse, "Detection of carbohydrates by capillary electrophoresis with pulsed amperometric detection", *Anal. Chem.*, 65 **1993** 948-951.

18. A. Klockow, A. Paulus, V. Figueiredo, R. Amado, and H.M. Widmer, "Determination of carbohydrates in fruit juices by liq6uid chromatography", *J. Chromatogr. A*, 680 **1994** 187-200.

19. W. Lu, and R.M. Cassidy, "Pulsed amperometric detection of carbohydrates separated by capillary electrophoresis", *Anal. Chem.*, 65 **1993** 2878-2881.

20. J. Ye, and R.P. Baldwin, "Amperometric detection in capillary electrophoresis with normal size electrodes", *Anal. Chem.*, 65 **1993** 3525-3527.

21. A.E. Bruno, B. Krattiger, F. Maystre, and H.M. Widmer, "On-column laser-based refractive index detector for capillary electrophoresis", *Anal. Chem.*, 63 **1991** 2689-2697.

22. A.E. Bruno, and B. Krattiger, "On-column refractive index detection of carbohydrates separated by HPLC and CE", J. Chromatogr. Libr, 58 **1995** 431-446.

23. J.M. Saz, B. Krattiger, A.E. Bruno, F. Maytsre, and H.M. Widmer, "Enhanced refractive index detection for capillary", *Anal. Methods Instrum.*, 1 **1994** 203-207.

24. S. Honda, S. Iwase, A. Makino, and S. Fukjimara, "Simultaneous determination of reducing monosaccharides by capillary zone elctrophoresis as the borate complexes of N-2-pyridylglycamines", *Anal. Biochem.*, 176 **1989** 72-77.

25. S. Honda, E. Akao, S. Suzuki, M. Okuda, K. Kakehi, and J. Nakamura,"HPLC of reducing carbohydrates as strongly ultraviolet-absorbing and electrochemically sensitive 1-phenyl-3-methyl-5-pyrazolone derivatives", *Anal. Biochem.*, 180 **1989** 351-357.

26. Y. Mechref, and Z. El Rassi, "Capillary zone electrophoresis of derivatized acidic monosaccharides", *Electrophoresis*, 15 **1994** 627-634.

27. J.Y. Zhao, P. Diedrich, Y. Zhang, O. Hindsgaul, and N.J. Dovichi, "Separation of aminated monosacchairdes by capillary zone electrophoresis with laser-induced fluorescence", *J. Chromatogr. B*, 657 **1994** 307-313.

28. L. Blomberg, J. Wieslander, and T. Norberg, "Immobilization of reducing oligosaccharides to matrices by glycosylation linkage*", J. Carbohydr. Chem.*, 12 **1993** 265-276.

29. R.F. Borch, M.D. Bernstein, and H.D. Durst, "Cyanohydridoborate anion as a selective reducing agent", *J. Am. Chem. Soc.*, 93 **1971** 2897-2904.

30. S. Hase, T. Ibuki, and T. Ikenaka, "Reexamination of the amination with 2-aminopyridine used for fluorescence labeling of oligosaccharides and ist applications to glycoproteins", *J. Biochem.*, 95 **1984** 197-205.

31. P. Jackson "The use of polyacrylamide-gel electrophoresis for the high-resolution separation of reducing saccharides labelled with 8-aminonaphthalene-1,3,6-trisulfonic acid", *Biochem. J.*, 270 **1990** 705-713.

32. C. Chiesa, and C. Horvath, "Capillary zone electrophoresis of malto-oligosaccharides derivatized with 8-aminonaphthalene-1,3,6,-trisulfonic acid", *J. Chromatogr.*, 645 **1993** 337-352.

33. A. Klockow, H.M. Widmer, R. Amadò, and A. Paulus, "Capillary electrophoresis of ANTS labeled oligosaccharide ladders and complex carbohydrates", *Fresenius' J. Anal. Chem.*, 350 **1994** 415-425.

34. A.D. Tran, S. Park, P.J. Lisi, O.T. Huynh, R.R. Ryall, and P.A. Lane, "Separation of carbohydrate-mediated microheterogeneity of recombinant human erythropoietin by free solution capillary additives", *J. Chromatogr.*, 542 **1991** 459-471.

35. A. Klockow, R. Amadò, H.M. Widmer, and A. Paulus, "Separation of 8-amino naphthalene-1,3,6,-trisulfonic acid-labeled nueral and sialylated N-linked complex oligosaccharides by capillary electrophoresis", *J. Chromatogr. A*, 716 **1995** 241-257.

36. E. Grill, C. Huber, P. Oefner, A. Vorndran, and G. Bonn, "Capillary zone elctrophoresis of p-aminobenzoic acid derivatives of aldoses, ketoses and uronic acids", *Electrophoresis*, 14 **1993** 1004-1010.

37. C. Huber, E. Grill, P. Oefner, and O. Bobleter, "Capillary electrophoretic determination of the component monosaccharides in hemicelluloses ", *Fresenius' J. Anal. Chem.*, 348 **1994** 825-831.

38. A.E. Vorndran, E. Grill, C. Huber, P.J. Oefner, and G.K. Bonn, "Capillary zone electrophoresis of aldoses, ketoses and uronic aids derivatized with ethyl-p-aminobenzoate", *Chromatographia*, 34 **1992** 109-114.

39. H. Schwaiger, P.J. Oefner, C. Huber, E. Grill, and G.K. Bonn, "Capillary zone electrophoresis and micellar electrokinetic chromatography of 4-aminobenzonitrile carbohydrate derivaties", *Electrophoresis*, 15 **1994** 941-952.

40. J. Liu, O. Shirota, D. Wiesler, and M. Novotny, "Ultrasensitive fluorometric detection of carbohydrates as derivatives in mixtures separated by capillary", *Proc. Natl. Acad. Sci. U. S. A.*, 88 **1991** 2302-2306.

41. J. Liu, O. Shirota, and M. Novotny, "Separation of fluorecent oligosaccharide derivatives by microcolumn technique based on electrophoresis and liquid chromatography", *J. Chromatogr.*, 559 **1991** 223-235.

42. J. Liu, O. Shirota, and M. Novotny, "Sensitive laser-assisted determination of complex oligosaccharide-mixtures separated by capillary gel electrophoresis at high resolution", *Anal. Chem.*, 64 **1992** 973-975.

43. J. Sudor, and M. Novotny, "Electromigration behavior of polysaccharides in capillary electrophoresis under pulsed-field conditions", *Proc. Natl. Acad. Sci. USA*, 90 **1993** 9451-9455.

44. K. Kakehi, S. Suzuki, S. Honda, and Y.C. Lee, "Precolumn labeling of reducing carbohydrates with 1-(p-methoxy)phenyl-3-methyl-5-pyrazolone: Analysis of neutral and sialic acid containing oligosaccharides found in glycoproteins", *Anal. Biochem.*, 199 **1991** 256-268.

45. S. Honda, T. Ueno, and K. Kakehi, "High-performance capillary electrophoresis of unsaturated oligosaccharides from glycosaminoglycans by digestion with chondroitinase ABC as 1-phenyl-3-methyl-5-pyrazolone derivatives", *J. Chromatogr.*, 608 **1992** 289-295.

46. S. Honda, K. Togashi, and A. Taga, "Unusual separation of 1-phenyl-3-methyl-5-pyrazolone derivatives of aldoses by capillary zone electrophroresis", *J. Chromatogr. A*, 791 **1997** 307-311.

47. M. Novotny, and J. Sudor, "High performance capillary electrophoresis of glycoconjugates", *Electrophoresis*, 14 **1993** 373-389.

48. E.F. Hounsell, "Glycoprotein analysis in Biomedicine", *Methods in Molecular Biology*, 14 **1993** 1-15.

49. W. Nashabeh, and Z. El Rassi, "Capillary zone electrophoresis of linear and branched oligosaccharides", *J. Chromatogr.*, 600 **1992** 279-287.

50. K.B. Lee, Y.S. Kim, and R.L. Linhardt, "Capillary zone electrohoresis for the quantitation of oligosaccharides formed through action of chitinase", *Electrophoresis*, 12 **1991** 636-640.

51. A. Paulus, and A. Klockow, "Detection of carbohydrates in capillary", *J. Chromatogr. A*, 720 **1996** 353-376.

52. G. Okafo, L. Burrow, S.A. Carr, G.D. Roberts, W. Johnson, and P. Camilleri, "A coordinated high-performance liquid chromatographic, capillary electrophoresis and mass spectrometric approach for the analysis of oligosaccharide mixtures derivatized with 2-aminoacridone", *Anal. Chem.*, 68 **1996** 4424-4430.

53. P. Camilleri, G.B. Harland, and G. Okafo, "High resolution and rapid analysis of branched oligosaccharides by capillary electrophoresis", *Anal. Biochem.*, 230 **1995** 115-122.

54. R.A. Evangelista, M.-S. Liu, and F.-T.A. Chen, "Characterization of 9-Aminopyrene-1,4,6-trisulfonate Derivatized Laser-Induced Fluorescence Detection", *Anal. Chem.*, 67 **1995** 2239-2245.

55. W. Nashabeh, and Z. El Rassi, "Capillary zone electrophoresis of pyridylamino derivatives of maltololigosaccharides", *J. Chromatogr.*, 514 **1990** 57-64.

56. P.D. Grossman, J.C. Colburn, and H.H. Lauer, "A semiempirical model for the electrophoretic mobilities of peptides in free-solution capillary electrophoresis", *Anal. Biochem.*, 179 **1989** 28-33.

57. C. Chiesa, and R.A. O'Neill, "Capillary zone electrophoresis of oligosaccharides derivatized with various aminonaphthalene sulfonic acids", *Electrophoresis*, 15 **1994** 1132-1140.

58. P.J. Oefner, and C. Chiesa, "Capillary electrophoresis of carbohydrates", *Glygobiology*, 4 **1994** 397-412.

59. A. Klockow, R. Amadò, H.M. Widmer, and A. Paulus, "The influence of buffer composition on separation efficiency and resolution in capillary electrophoresis of 8-aminonaphthalene-1,3,6-trisulfonic acid labeled monosaccharides and complex carbohydrates", *Electrophoresis*, 17 **1996** 110-119.

60. S. Honda, A. Makino, S. Suzuki, and K. Kakehi, "Analysis of the oligosaccharides in ovalbumin by high-performance capillary electrophoresis", *Anal. Biochem.*, 191 **1990** 228-234.

61. S. Suzuki, K. Kakehi, and S. Honda, "Two-dimensional mapping of N-glycosidically linked asia-lo-oligosaccharides form glycoproteins as reductivley pyridylaminated derivatives using dual separation modes of high performance capillary electrophoresis", *Anal. Biochem.*, 205 **1992** 227-236.

62. J.P. Landers, R.P. Oda, B.J. Madden, and T.C. Spelsberg, "High-performance capillary electrophoresis of glycoproteins: the use of modifiers of electroosmotic flow for analysis of microheterogeneity", *Anal. Biochem.*, 205 **1992** 115-124.

63. M. Stefansson, and M. Novotny, "Modification of the electrophoretic mobility of neutral and charged polysaccharides", *Anal. Chem.*, 66 **1994** 3466-3471.

64. M. Stefansson, and M. Novotny, "Resolution of the branched forms of oligosaccharides by high performance capillary electrophoresis", *Carbohy. Res.*, 258 **1994** 1-9.

65. S. Honda, K. Yamamoto, S. Suzuki, M. Ueda, and K. Kakehi, "High performance capillary zone electrophoresis of carbohydrates in the presence of alkalline earth metal cations", *J. Chromatogr.*, 588 **1991** 327-336.

66. G.N. Okafo, L.M. Burrow, W. Neville, A. Truneh, R.A.G. Smith, M. Reff, and P. Camilleri, "Simple differentiation between core-fucosylated and nonfucosylated glycans by capillary electrophoresis", *Anal. Biochem.*, 240 **1996** 68-74.

67. A. Guttman, and C. Starr, "Capillary and slab gel electrophoresis profiling of oligosaccharides", *Electrophoresis*, 16 **1995** 993-997.

68. A. Guttman, and T. Pritchett, "Capillary gel electrophoresis separation of high-mannose type oligosaccharides derivatized by 1-aminopyrene-3,6,8-trisulfonic acid", *Electrophoresis*, 16 **1995** 1906-1911.

69. A. Guttman, F.-T.A. Chen, R.A. Evangelista, and N. Cooke, "High-resolution capillary gel electrophoresis of reducing oligosaccharides with 1-aminopyrene-3,6,8-trisulfonic acid", *Anal. Biochem.*, 233 **1996** 234-242.

70. A. Guttman, "High-resolution carbohydrate profinling by capillary gel electorphoresis", *Nature*, 380 **1996** 461-462.

71. A. Guttman, F.-T.A. Chen, and R.A. Evangelista, "Separation of 1-aminopyrene-3,6,8-trisulfonate-labeled asparagine-linked fetuin glycans by capillary gel electrophoresis", *Electrophoresis*, 17 **1996** 412-417.

72. A. Guttman, "Multistructure sequencing of N-linked fetuin glycans by capillary gel electrophoresis and enzyme matrix digestion", *Electrophoresis*, 18 **1997** 1136-1141.

73. A. Guttmann, and K. Williams Ulfelder, "Exoglycosidase matrix mediated sequencing of a complex glycan pool by capillary electrophoresis" *J. Chromatogr.* A, 781 **1997** 547-554.

5 Applications

5.1 Mono- and disaccharides

The use of CE for the analysis of carbohydrates was first demonstrated for separations of standard mixtures of mono-and disaccharides in order to develop suitable CE separation systems and detection schemes for this class of compounds. Since the different carbohydrate separation and detection strategies have already been reviewed in Chapter 4, this section will focus on applications. There are three main areas in which mono- and disaccharide analysis is of importance: food analysis, compositional analysis of oligo- and polysaccharides and carbohydrate analysis in biological samples such as urine or blood.

CE with indirect UV detection using sorbic acid as the chromophore was demonstrated to be a simple and inexpensive method for the determination of soluble low-molecular weight carbohydrates in a number of fruit juices [1, 2]. The suitability of this method for carbohydrate analysis was further enhanced by comparison with a routine high performance anion exchange chromatographic (HPAEC) method [1]. Provided a carbohydrate concentration in the low millimolar range, indirect UV detection is also applicable to other food samples than fruit juices, as shown by Lee and Lin [3] for 20 different samples. These samples can be categorized into sport drinks, nutrient added drinks, natural or artificial flavor fruit juices and dairy products. Analysis of carbohydrates in food samples can also be performed using amperometric detection. After separation in a sodium hydroxide electrolyte saccharose, lactose, galactose, and glucose were detected in a chocolate beverage [4]. The same separation and detection approach was applied to the analysis of monosaccharides in apple juice as well as for monitoring the activity of the enzyme glucose oxidase which oxidizes glucose to form gluconic acid and hydrogen peroxide [5]. Reductive amination of ketoses results in the introduction of a new chirality center and thus in the formation of two diastereomers per ketose enantiomer, as shown in Figure 5.1 [6]. Consequently, three peaks showed up in the electropherogram of a honey extract derivatized with (S)-(–)-1-phenylethylamine, one for glucose and two for the D-fructose diastereomers (Figure 5.2).

Composition analysis of oligo- and polysaccharides advanced to become an important application area in CE carbohydrate analysis. Huber et. al [7] determined the composition of hemicelluloses isolated from a number of plants, after their hydrolysis with trifluoracetic acid and derivatization of the released monosaccharide units with p-aminobenzoic acid. The major component of the investigated hemicelluloses was found to be xylose with relative amounts between 64.5% and 80%. Next to the neutral monosaccharides Glc, Gal, Ara, Fuc also small amounts of GlcA and GalA were found. The suitability of CE for composition analysis of glycoproteins could be demonstrated as well. By means of acid hydrolysis mono-

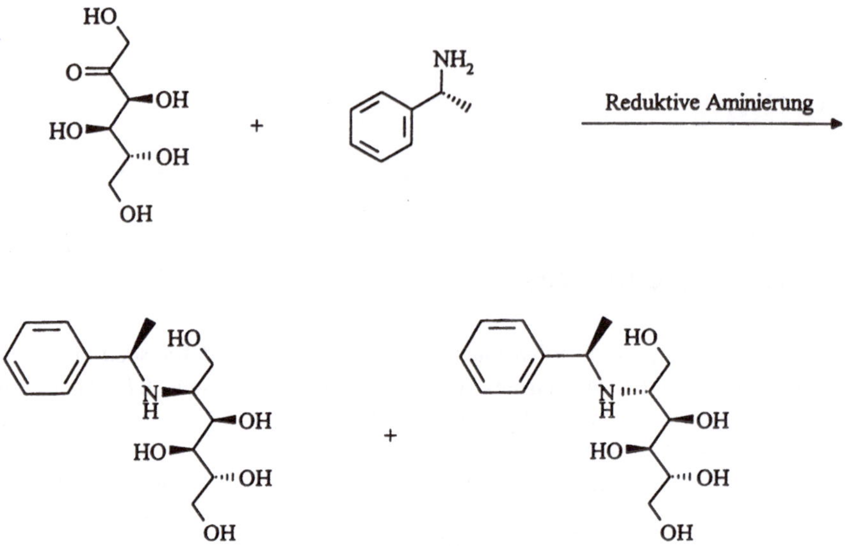

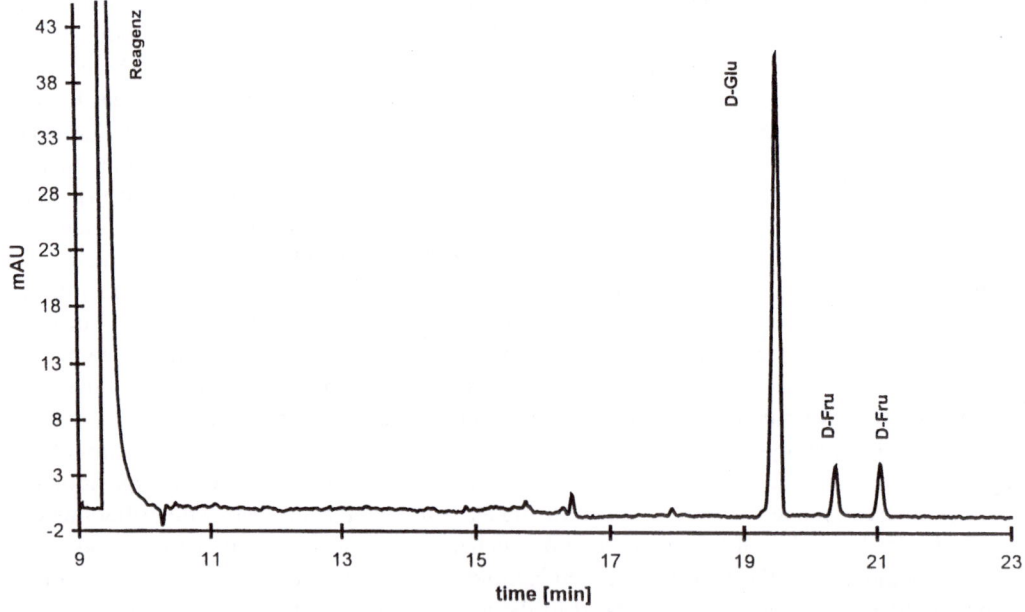

Figure 5.1   Reductive amination of fructose (from [6], with permission)

Figure 5.2   Separation of the derivatives obtained by reductive amination of a honey extract. Separation conditions: BGE, 50 mM borate, pH 10.3; capillary, 87/80 cm x 50 μm ID; voltage, 25 kV; temperature, 20°C; UV detection, 200 nm (from [6], with permission).

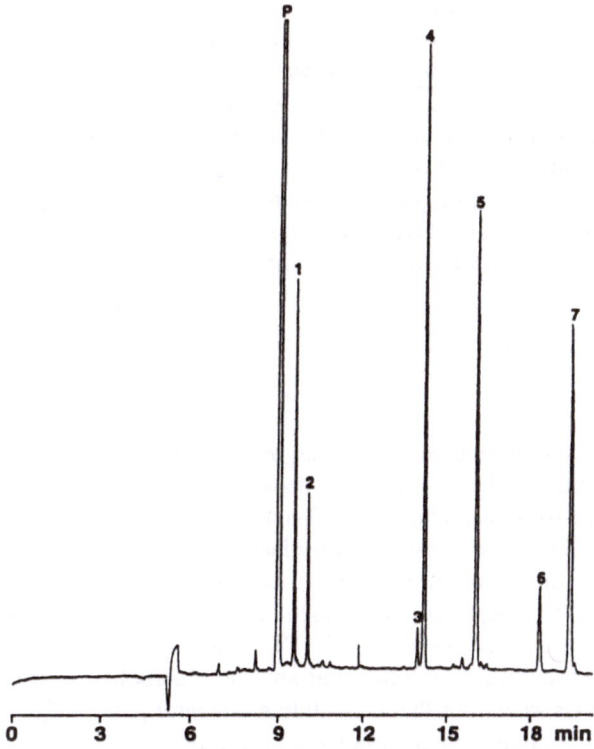

Figure 5.3 MEKC of PMP-derivatized monosaccharides derived from fetal calf serum fetuin. Separation conditions: BGE, 30 mM Tris-phosphate, pH 7.5, 50 mM SDS; capillary, 72/50 cm x 50 µm ID; voltage, 20 kV (~ 29 µA); temperature, 30°C; UV detection, 245 nm. Peak assignment: P = PMP, 1 = Man, 2 = talose (internal standard), 3 = Glc (contaminant), 4 = glucosamine, 5 = Gal, 6 = galactosamine, 7 = 2-deoxy-D-ribose (reference standard) (from [8], with permission).

saccharides were released from bovine fetuin and separated after derivatization with PMP [8] or CBQCA [9], respectively. Figure 5.3 shows that the major monosaccharides found in fetuin are Man, GlcNAc, Gal and GalNAc with molar ratios of 3.0:4.4:3.9:0.6 [8]. Unlike neutral and amino sugars, sialic acid does not require a charged label for electrophoretic separation due to its intrinsic negative charge. Derivatization with both a charged and a neutral label allows an easy differentiation of neutral sugars, amino sugars and sialic acid. For example, the acid hydrolysis mixture of bovine fetuin could be characterized this way [10]. In a similar manner the monosaccharide components of various di- and oligosaccharides such as lactose, melobiose, rutin, digitonin and arabic gum were quantified [11].

The potential use of CE for analysis of carbohydrates in biological samples was proved by determination of glucose levels in human blood [12, 13] by pulsed amperometric detection. The only sample preparation steps required were centrifugation and filtration. Quantitation of glucose in blood is a valuable tool in the diagnosis of medical disorders such as diabetes. Amperometric detection was also used to detect electrophoretically separated mono- and disaccharides in urine samples [4]. After dilution with the BGE and filtering prior to injection, saccharose, lactose, glucose and fructose were identified in the urine sample. Application of CE to sample volumes smaller than 500 nl was demonstrated by the determination of glucose in tear samples [14]. Two enzymatic reactions resulting in a fluorescent product allowed non-invasive monitoring of glucose levels in 200 nl sample volumes. In the first step, glucose is oxidized in the presence of glucose oxidase producing hydrogen peroxide, which reacts quantitatively with the fluorogenic compound homovanillic acid. The second reaction is catalyzed by the enzyme peroxidase. Using this enzymatic approach the concentration of glucose in human tear samples was found to be approximately 138 µM.

5.2 Oligosaccharides

The differentiation of oligo-and polysaccharides is not consistent in the published literature. Per definition, oligosaccharides are carbohydrates with less than 10 defined monomeric structures, each with a defined linkage to its neighbors (see Chapter 3). However, in most cases the term oligosaccharide is applied to both small and very large oligomers, consisting of up to 30 and more monosaccharide units. Additionally, high molecular weight polysaccharides are usually analyzed through their degradation products, obtained by chemical or enzymatic hydrolysis, which in many cases are oligosaccharides. Therefore, in this case it is also preferred to apply the term oligosaccharide to both small and large carbohydrate oligomers. For clarity, the oligo- and polysaccharides that constitute an integral part of glycoproteins and proteoglycans are treated separately in the following paragraphs.

5.2.1 Underivatized oligosaccharides

Underivatized oligosaccharides can be detected either by UV detection at low wavelengths of 195 nm or through amperometry. The α-galactosides stachyose, verbascose and ajugose are synthesized in various plants through successive addition of galactosyl residues to raffinose. Quantitative CE determination of these raffinose-oligosaccharides by borate complexation and detection at 195 nm, allowed for the characterization of the quality and the nutritive value of the legume seeds from which they were isolated [15]. Direct UV detection at 185 nm was performed for the analysis of colominic acids which are a mixture of α-2,8-linked homopolymers of NeuAc with varying degree of polymerization (dp). Using an alkaline borate electrolyte containing SDS and methylcellulose, baseline separation up to a dp of 30 was achieved [15a]. The same separation system was applied to monitor the hydrolysis of the NeuAc-polymers by neuraminidase treatment.

Recently, Zhou and Baldwin [16] presented an analysis scheme for oligo- and polysaccharides using constant-potential amperometric detection at copper electrodes. Separation of the larger oligomers with 100 mM NaOH as the background electrolyte required reversal of the EOF through addition of a cationic surfactant in order to increase the electromigration rate of the anionic oligosaccharides. Reversed EOF and reversed polarity resulted in the same direction for the electrophorectic mobility of the oligosaccharide anions and the EOF, thus allowing resolution of standard maltooligomers and starch hydrolyzates up to dp > 10. One drawback of this approach was found to be the non-linearity of the detector response. The detector response depends on the number of hydroxyl groups which are in contact with the electrode surface which again is a function of the number of hydroxyl groups in the carbohydrate molecule and its three dimensional structure.

5.2.2 Derivatized oligosaccharides

During the last few years a considerable number of CE separations of labeled polysaccharide derived oligosaccharides have been published. Table 5.1 gives a comprehensive overview over the different applications.

The first papers discussing CE analysis of derivatized oligosaccharides demonstrated the separation of 2-aminopyridine [17] and CBQCA [9] labeled maltooligosaccharides as well as the separation of PMP-derivatives of laminaro-, isomalto- and cello-oligosaccharides [18]. The oligosaccharide derivatives, except the aminopyridine conjugates, were separated through borate complexation. The borate complexes migrate against the EOF in the order of decreasing size. Because of the unfavorable charge-to-mass ratio (q/m) at higher dp's the resolution decreases with increasing number of monosaccharide units. This limitation can be overcome by separating the oligosaccharides under 'no-EOF' (arheic) conditions. Using a pH 2.5 phosphate BGE and reversed polarity, Chiesa and Horváth [19] were successful in resolving negatively charged ANTS labeled maltooligosaccharides up to dp 30 with excellent resolution (see Figure 4.13). Almost the same separation conditions were applied for the analysis of the carbohydrate fraction extracted from 'Gummibärchen' [20], which mainly consists of saccharose and starch syrup (acid hydrolyzed starch). Next to the linear maltooligosaccharides presumedly branched isomers showed up in the electropherogram as small peaks preceding the main signals (Figure 5.4). A second approach to achieve high resolution of high molecular weight oligosaccharides is the use of coated capillaries with no apparent EOF. By means of polyacrylamide coated capillaries and a Tris-borate buffer resolution of ANTS derivatized dextran [21] and corn amylose [22] oligomers up to more than a dp of 60 was achieved.

The linear relationship between electrophoretic mobility and molecular mass or number of glucose molecules, respectively, found for ANTS labeled maltooligosaccharides (see Figure 4.14 [19, 23]) was also proved for dextran oligomers labeled with N-(4-aminobenzoyl)-L-glutamic acid [24] and for CBQCA derivatives separated in gel filled capillaries [19]. The use of oligosaccharide homologues offers a convenient means of testing electrophoretic systems in the light of these linear relationships and may help to identify unknown compounds in complex oligosaccharide mixtures.

The effect of structure and charge of the derivatizing agent on resolution and separation efficiency was investigated for various naphthalene sulfonic acid based derivatizing agents such as ANTS, ANDSA and different ANSA's [21, 23]. Using derivatized maltooligosaccharides, dextrins or dextran hydrolyzates as model mixtures it was shown that both resolution and separation efficiency as well as the migration time heavily depend on the number of negatively charged sulfonate groups in the fluorescent label with ANTS giving the best

Table 5.1 CE applications of polysaccharide derived derivatized oligosaccharides

Oligosaccharides derived from	Label	Separation conditions	Reference
Amylose	ANTS	C: PA-coated; B: 200 mM Tris-borate, pH 8.65	[22]
Amylopectin	ANTS	C: PA-coated; B: 200 mM Tris-borate, pH 8.65	[22]
k-Carrageenan	ANTS, 6-AQ	C: PA-coated; B: 25 mM sodium citrate, pH 3.0 C: PA-coated; B: 25 mM sodium citrate, pH 3.0 + 1% linear PA	[25]
Cellulose	PMP	C: uncoated; B: 200 mM borate, pH 9.5	[18]
Chitin	2-AP, 6-AQ ANDSA	C: polyether interlocked coating; B: 100 mM phosphate + TTAB, pH 5.0 C: uncoated; B:50 mM boric acid, pH 8.8	[27] [35]
Chitosan	CBQCA	C: uncoated; B: 10 mM phosphate/30 mM borate/50 mM SDS, pH 9.4	[166]
Dextran	ANTS ANTS APG	C: uncoated; B:50 mM phosphate, pH 2.5 C: PA-coated; B:100 mM Tris-borate, pH 8.65 C: hydroxypropylcellulose coated; B: 100 mM Tris-borate, pH 8.8	[20] [21] [24]
(Poly-)-Isomaltose	PMP 2-AP	C: uncoated; B: 200 mM borate, pH 9.5 C: PA-coated; B: 100 mM phosphate, pH 2.5	[18] [60]
Laminarin	ANTS PMP	C: PA-coated; B: 100 mM Tris-borate, pH 8.65 C: uncoated; B: 200 mM borate, pH 9.5	[21] [18]
Lichenan	ANTS	C: PA-coated; B: 200 mM Tris-borate, pH 8.65	[22]
Polygalacturonic acid	ANTS 2-AP	C: uncoated; B: 50 mM phosphate, pH 2.5 C: polyether coating with switchable EOF; B: 100 mM phosphate, pH 6.5	[20] [167]
Pullulan	ANTS	C: PA-coated; B: 200 mM Tris-borate, pH 8.65	[22]
Starch (Maltooligo-saccharides)	ANTS, ANDSA, ANSA 2-AP CBQCA	C: uncoated ; B: phosphate, pH 2.5 (+ TEA) C: uncoated; B: 100 mM phosphate, pH 3–5.0 C: uncoated; B: phosphate/borate	[20] [23] [17] [9]
Xyloglucans	2-AP 6-AQ	C: polyether interlocked coating; B: 100 mM phosphate + 50 mM TTAB, pH 4.75 C: uncoated, B: 420 mM borate, pH 9.0	[27] [28, 30]

ANTS: aminonaphthalenetrisulfonic acid, ANDSA: aminonaphthalenedisulfonic acid, ANSA: amino-naphthalenesulfonic acid, 6-AQ: 6-aminoquinoline, 2-AP: 2-aminopyridine, CBQCA: 3-(4-carboxy-benzoyl)-2-quinolinecarboxyaldehyde, PMP: 1-phenyl-3-methyl-2-pyrazolin-5-one, APG: N-(4-ami-nobenzoyl)-L-glutamic acid, C: Capillary, B: Buffer, PA: polyacrylamide, SDS: sodium dodecylsulfa-te, TEA: triethylamin, TTAB: Tetradecyltrimethylammonium bromide

results. Comparing the migration behavior of κ-carrageenan hydrolyzates labeled with different tags revealed that the migration depends on the influence the label has on the charge-to-friction ratio of the labeled molecules [25]. Unlike the charge-to-mass ratio, the charge-to-friction ratio also includes the impact of the conformational structure of the solute on its

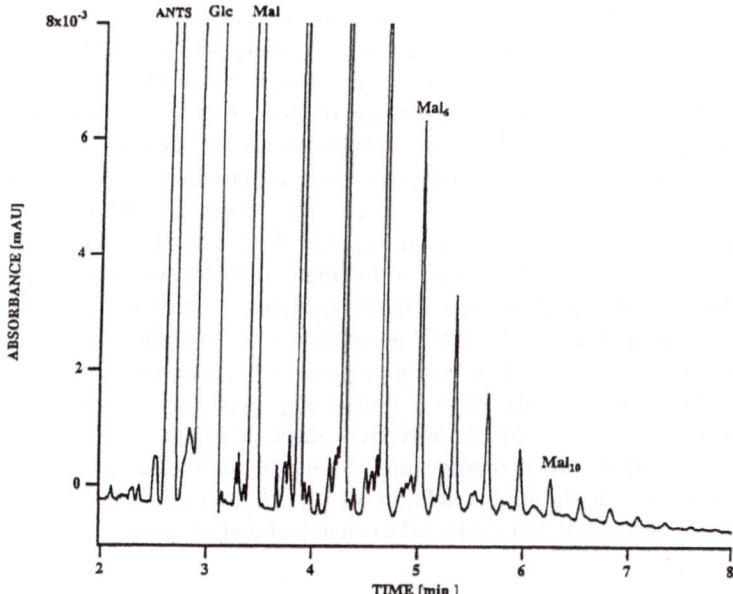

Figure 5.4 Separation of the ANTS-derivatized carbohydrate fraction, isolated from 'Gummibär-chen'. Separation conditions: BGE, 50 mM phosphate, pH 2.5; capillary, 42/35 cm x 50 μm ID; voltage, 20 kV; temperature, 25°C; injection, 10 sec with 53 mbar; UV detection, 223 nm (from [20], with permission).

migration behavior. Since ANTS increases the charge-to-friction ratio of the oligosaccharides, the migration order reverses from small to the large. With 6-aminoquinoline where the charge-to-friction ratio is decreased, the migration order is entirely reversed. The applications discussed in this paragraph again demonstrate the importance of zero EOF and the use of a sieving media on the resolution of oligomers with higher dp [25].

Guttman et al. [26] applied the APTS labeling strategy to food analysis by profiling the carbohydrate pattern in different beer samples. Separation of the labeled carbohydrate fraction in a replaceable polymer network matrix revealed comparable patterns for the different beer types but with variable amounts of the individual carbohydrates. The beer carbohydrate fingerprints were compared to the maltooligosaccharide ladder where the individual peaks can be assigned to distinct dp's. That way the size range of the carbohydrates detected in the different beer types could be estimated. The authors stated that with the increasing use of novel brewing adjuncts and carbohydrate sources, the monitoring of individual sugars might be useful for determining metabolic characteristics and potential fermentability in the brewing process.

The high resolution potential of CE for linear oligosaccharides was also exploited in the separation of branched hetero-oligosaccharides derived from large xyloglucan polysaccharides of cotton cell walls by cellulase digestion [27]. The variations in the xyloglucans are caused by differences in the nature and distribution of the side chains attached to the β-1,4-glucose backbone. Cellulase digests the backbone after any glucosyl residue that does not contain a side chain thus liberating fragments of the polymer which reflect its branching pattern. The pyridylamino (2-AP) derivatized xyloglucan fragments of a cellulase digest, have all the same charge and therefore migrated in the order of increasing size and decreasing charge-to-mass ratio (Figure 5.5). However, for the same number of residues with only slight differences in the molecular weight, the less branched oligosaccharides eluted earlier than the more branched ones. Based on this observation Nashabeh and El Rassi [27] developed a mobility indexing system for branched xyloglucan oligosaccharides with respect to the mobility of the pyridylamino-derivatives of linear N-acetylchito-oligosaccharides. Endoxylanase digestion followed by CE of birch and pine kraft pulps allowed for the characterization of xylooligosaccharide mixtures with respect to distribution, molecular weight and composition [28, 29]. Whilst the unsaturated acidic oligosaccharides, ranging in size from the tri- to nonasaccharides, were detected by their UV absorption at 232 nm the neutral xylooligosaccharides were derivatized with 6-aminoquinoline prior to CE analysis. This study revealed that the resolution of structural analogs was greatly improved by converting the unsaturated xylooligosaccharides to their alditol derivatives through sodium borohydride reduction. The 6-aminoquinoline derivatization procedure was also applied to the analysis of neutral β-1,4-D-xylooligomers and acidic oligosaccharides [30]. The xylooligosaccharides were isolated from birch or spruce wood and subjected to chemical and enzymatic hydrolysis prior to the derivatization process.

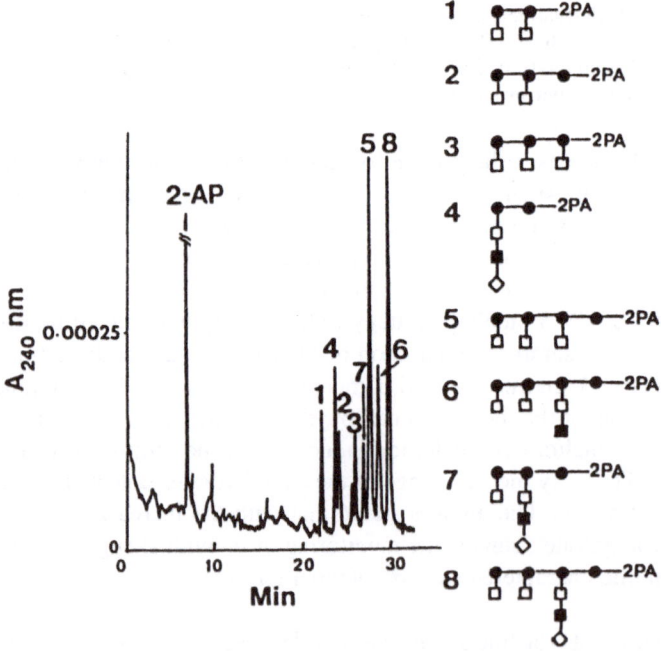

Figure 5.5   Capillary zone electrophoresis mapping of pyridylamino derivatives of xyloglucan oligosaccharides from cotton cell walls. Separation conditions: BGE, 0.1 M sodium phosphate containing 50 mM tetrabutylammonium bromide, pH 4.75; capillary, 80/50 cm x 50 μm ID; voltage, 20 kV. Symbols: 2-AP = 2-aminopyridine, ● = Glc, □ = Xyl, ■ = Gal, ◇ = Fuc (from [27], with permission).

A special group of oligosaccharides are the cyclodextrins (CD) which are neutral cyclic polymers of glucose (see Chapter 3.2). They are used for example to increase the solubility and bioavailability of hydrophobic pharmaceuticals in aqueous solution and as selectivity agents for the resolution of structural, positional and enantiomeric isomers in analytical chemistry. One approach to analyze the neutral CD's by CE is indirect UV detection [31]. In this case benzoic acid was used to form an inclusion complex with the CD's thus assuring their electromigration, and to serve as the UV absorbing constituent of the BGE. The selectivity was obtained through the inclusion complex equilibria of the different CD's and the benzoic acid. Enhanced resolution of the α, β, and γ-CD's by adjusting accurately the pH of the BGE and the benzoic acid concentration allowed for the analysis of CD's in complex matrices such as fermentation broth, urine, plasma and pharmaceutical formulations [32]. A similar dynamic labeling approach was applied by Penn et al. [33]. They used the fluorescent 2-anilinonaphthalene-6-sulfonic acid (2,6-ANS) for visualization and for imparting a charge to the neutral CD's through formation of inclusion complexes. The study revealed that depending on the nature of the CD the increase in fluorescence emission due to CD complexation of 2,6-ANS differed significantly. As indicated in Table 5.2, fluorescence emission was enhanced only by a factor of 1.8 for the α-CD, while it was enhanced a factor of 245 for the heptakis 2,6-di-O-methyl-β-CD, compared to the uncomplexed 2,6-ANS. Using this technique, components in 2,6-di-O-methyl-β-CD preparations, with differing degrees of substitution, were separated. Other charged background chromophores applied to the dynamic labeling of CD's are benzylamine, salicylic acid, sorbic acid and 1-naphthylacetic acid [34]. Depending on the pH of the BGE the migration of the CD's was based either on both the charges aquired as a result of the ionization of the CD hydroxyl groups and the charge aquired by the inclusion complex formed between the CD's and the background chromophore or only on the latter one. Analysis of derivatized CD's might be of increasing interest in the pharmaceutical industry since their intended use as drug delivery complexes will require the monitoring of their chemical composition as a critical part of any toxicological study [33].

5.2.3 Enzyme action on labeled oligosaccharides

Another field of carbohydrate analysis where CE proved to be useful is the monitoring of enzyme action on oligosaccharides in order to investigate both, the specificity of the enzyme and the structural features of the substrate oligosaccharide. Oligosaccharide mapping of laminarin before and after treatment with different β-1,3-glucose-hydrolyzing enzymes revealed the different specificity of these enzymes for the branched polysaccharide. Cellulase and laminarase caused more complete hydrolysis compared to the two endoglucanases EGI and EGII [21]. After selective debranching of ANTS labeled corn- and potato-amylopectins with isoamylase the branched species could be separated and detected next to the linear degradation products [22]. Comparison of the oligosaccharide maps of the two isoamylolysates proved a species-specific and polymodal distribution for linear and branched chain length. In the same paper the polydispersity of a lichenan sample (for structure see Table 3.3) was investigated by hydrolysis with laminarase. While before the enzymatic treatment

Table 5.2 Fluorescence signal properties of cyclodextrin-2,6-ANS complexes [33]

	$\lambda_{excitation\ max.}$	$\lambda_{emission\ max.}$	Fluorescence emission enhancement
uncomplexed 2,6-ANS	318 nm	464 nm	–
α-CD-2,6-ANS	318 nm	452 nm	1.8 (416 nm)
β-CD-2,6-ANS	317 nm	441 nm	55 (422 nm)
γ-CD-2,6-ANS	318 nm	454 nm	2.3 (426 nm)
Heptakis (2,6-di-O-methyl)-β-CD-2,6-ANS	317 nm	437 nm	245 (426 nm)

the polydisperse lichenan oligomers migrated in broad peaks they appeared to migrate in 'clusters' after enzymatic hydrolysis, as shown in Figure 5.6. The distance between each 'cluster' is approximately equivalent to an increase in chain length by 3-4 glucose units thus mirroring the ratio of maltotriose and maltotetraose units in the polysaccharide.

Lee, Kim and Linhardt [35] analyzed the products formed by chitinase acting on fluorescent labeled N-acetylchitooligosaccharides with dp 1–6. Exhaustive treatment of the hexasaccharide with the chitinase resulted in the formation of a mixture containing 8% tri, 80% di and 12%-monomer. It was concluded that the exo-type chitinase used in this study removes two monosaccharide residues at one time from the substrates non reducing termini. The absence of significant amounts of monomer after treatment of the disaccharide provided additional evidence to that explanation.

With polygalacturonic acid as a model substrate the enzymatic degradation process through pectinases was kinetically monitored [20]. In contrast to the former application the substrate was first incubated with the pectinase mix, which mainly contained endopoly-

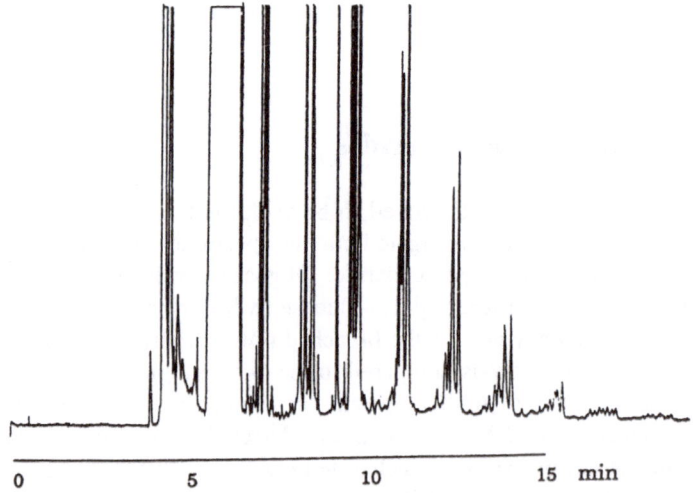

Figure 5.6 Oligosaccharide map of lichenan after laminarase treatment. Separation conditions: BGE, 0.2 M Tris-borate, pH 8.65; capillary, 35 cm (effective length) x 50 µm ID, polyacrylamide coated; voltage, –500 V/cm; LIF detection, excitation: 325 nm, emission: 518 nm (from [22], with permission).

galacturonase. The resulting oligosaccharide fragments were subsequently derivatized with ANTS and separated by CE. As shown in Figure 5.7, at the beginning of the degradation process, an accumulation of higher oligosaccharides (peak 7 and peak 8) was observed which seemed to be further degraded during the enzymatic hydrolysis. Smaller oligosaccharide fragments (peak 4) accumulated at the end of the process due to the higher affinity of the pectinases to larger oligomers. Peak 4, 7 and 8 in Figure 5.7 represent the presumed 4, 7 and 8-mer, respectively. Another assay to characterize pectate-depolymerizing enzymes used galacturonic acid polymers which were fluorescently end-labeled prior to enzymatic hydrolysis [36, 37]. This assay provided a way to examine the action of endo- and exo-polygalacturonases *in vitro* and *in vivo*. The *in vitro* studies revealed the different specificities of fungal and bacterial endopolygalacturonase. Whilst the active side of the enzyme produced from *A. niger* needs four GalA residues towards the non-reducing end the one from *E. carotovora* recognizes five residues. Acting on a GalA heptamer the fungal enzyme produced the labeled dimer, trimer and tetramer, whereas the bacterial enzyme produced only labeled trimer and tetramer. Isolation of the poly-GalA immediately after injection into cotyledons demonstrated the high exogalacturonase activity in the intercellular spaces of the cotyledons, since a big part of the hexamer injected was already converted into the pentamer [36]. Similar studies were performed using various di- and oligosaccharides as well as a biantennary asialo N-glycan as the substrates for β-galactosidases from *A. oryzae*, *E. coli*, *S. pneumoniae* and *Canavalia ensiformis* [38]. CE analysis of the different enzymatic digests after derivatization with ethyl-4-aminobenzoate again revealed different activities for the different β-galactosidases.

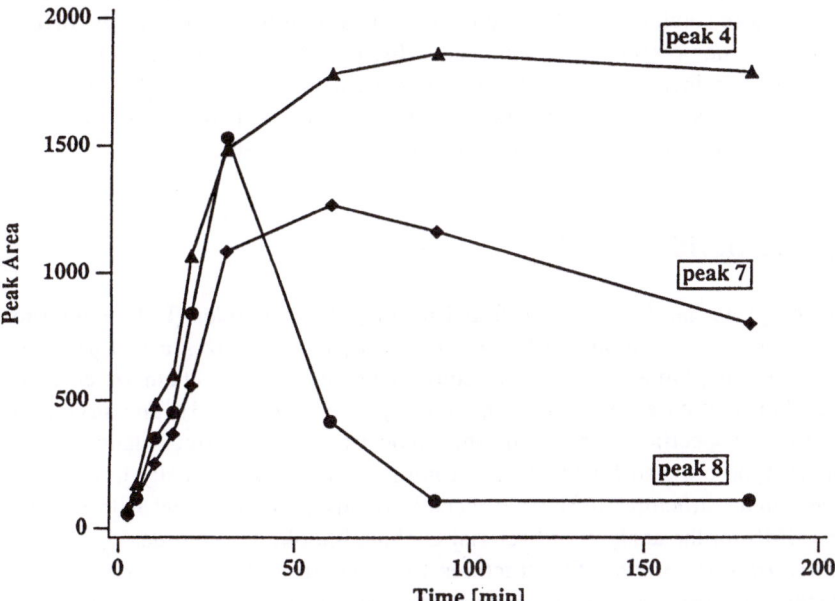

Figure 5.7 Formation and degradation of polygalacturonic acid oligomers as a function of incubation time. The oligomers were formed during the incubation of polygalacturonic acid with a pectinase preparation (from [20], with permission).

The TRSE-derivatization scheme (see chapter 4.2.1.4) was applied to monitor the enzyme products formed during incubation of yeast cells with the trisaccharide α-D-Glc-1,2-α-D-Glc-1,3-α-D-Glc-O-TMR, with O-TMR (Tetramethylrhodamine) being the fluorescent linker arm [39]. After 5 hours of incubation with the trisaccharide the formed yeast spheroblasts were lysed and injected into the CE system. The components were separated and detected by LIF. The observed conversion of the trisaccharide into the linker arm and intermediate di- and monosaccharides resulted from the sequential activity of α–glucosidase I+II present in yeast cells. These α-glucosidases are known to act specifically on α-D-Glc-1,2- and α-D-Glc-1,3-residues. The same system was used to monitor biosynthetic transformation of TRSE labeled N-acetyllactosamine (β-Gal-1,4-β-GlcNAc) in crude microsomal extracts [40]. Two different biosynthetic transformation were monitored: the α-1,2- and α-1,3- fucosylation of the N-acetyllactosamine through action of a fucosyl-transferase and the degradation of the lactosamine in to labeled β-GlcNAc and linker arm by galactosidase and hexosaminidase. In another application fucosyltranferase activity was assayed using enzymatically galactosylated chitobiose as the substrate [41].

An improved separation system was used to study isomeric oligosaccharides of the βGal-1,3-βGlcNAc and the βGal-1,4-βGlcNAc series after TRSE derivatization [42]. The TRSE labeled substrates and their potential enzymatic products were baseline resolved in a tetraborate/phenylboronic acid buffer containing SDS. Addition of phenylboronic acid to the separation buffer enhanced the selectivity and the resolution of the TRSE conjugates for two reasons. First, the relatively large size of the benzene group introduces additional steric effects on the formation of borate complexes thus enhancing the electrophoretic mobility differences among the oligosaccharides. Second, the phenylborate esters are more hydrophobic and therefore more soluble in the SDS micelles. The aim of this study was the investigation of enzyme activities present in mammalian cells. For that purpose human epidermoid carcinoma cells were incubated with the fluorescent substrate βGal-1,3-βGlcNAc-O-TMR. The CE-LIF analyses showed formation of both synthetic and hydrolytic products suggesting the action of glycosyltransferases and glycosidases in the cells.

5.2.4 Lipooligosaccharides

A special class of oligosaccharides are those found in lipopolysaccharides (LPS) bound to the surface of gram negative bacteria. Each bacterial serotype synthesizes a unique LPS, characterized by a specific composition and structure of the so called O-chain which consists of hexose residues and by an individual antigenicity (see Chapter 3.3). The structural characterization of the O-specific chains is not only important for the correct identification of the bacterium serotype but also for the production of antibodies specifically targeted to those immunodeterminant structures [43]. Two recently published papers deal with the application of CZE-ESMS for the analysis of bacterial O-chain lipooligosaccharides (LOS) derived from *Yersinia ruckeri* serotype O₂ structures [43] and from *Moraxella catarrhalis* [44]. While *Y. ruckeri* has been recognized as the causative agent of enteric redmouth disease in salmon fish, thus accounting for significant economic losses in aquaculture industry, *M. catarrhalis* was found to be an important human pathogen in many cases of upper respiratory tract and middle ear infections. Separation of LOS arising from mild hydrazinolysis of the intact LPS was achieved in aqueous ammonium formate BGE's. This allowed identification of sites of heterogeneity such as phosphate, phosphoethanolamine and acyl groups in case of *M. catarrhalis* or phosphate, acetyl and heptose residues in case of *Y. ruckeri,* on either the lipid A or the core oligosaccharides. More complex mixtures, e.g. resulting from complete deacetylation and dephosphorylation of LOS, were analyzed using anionic (ammonium formate, pH 8.5) and cationic separation conditions (2 M formic acid, polybrene coated capillary, reversed polarity) which are compatible with the operation conditions of the ESMS. Structural characterization of LOS and Lipid A disaccharides obtained from hydrazinolysis and acid hydrolysis of the intact LPS, was performed by tandem MS after introduction of the samples by direct flow injection. The authors stated that CE-ESMS provided an efficient analytical tool for profiling relative intact forms of bacterial endotoxins.

5.3 High molecular weight polysaccharides

So far only a few attempts have been made for the application of CE to intact polysaccharides. In the first paper dealing with polysaccharide separations in open tube capillaries a system consisting of a 1 mM fluorescein BGE of pH 11.5 was applied [45]. This alkaline fluorescein BGE assured the partial ionization of the analytes and allowed to perform indirect fluorescence detection. Nine different polysaccharides could be detected in this system. Since the migration time range for the polysaccharides is not as large as the range for the small mono- and disaccharides only a number of polysaccharide mixtures could be separated, for example a mixture of dextran, comb-dextran, hydroxyethylamylose and amylose.

The separation of CBQCA labeled polysaccharides such as chitosan, dextran or water-soluble cellulose-derivatives was also difficult to achieve since borate complexation resulted in only minor mobility differences and thus low resolution. The use of a sieving matrix was also not successful because due to their large molecular size the polysaccharides did not penetrate the gel network [46]. These problems were overcome by applying a potential gradient, which was inverted at a 180° angle, along the separation capillary, filled with an entangled polymer solution. In contrast to the constant potential conditions with no size separation due to molecular stretching of the polysaccharides, the molecules undergo shape transition under pulsed-field condition leading to a reptation like motion of the polysaccharides through the gel, according to their molecular size [46]. Under these optimized conditions separation of CBQCA labeled polydextrans of molecular weights in the range of 8'000 –2'000'000 was achieved.

Dextrans of lower molecular weight (M_r = 1'500–18'300) were separated under highly alkaline conditions with reversed EOF and electrochemically detected at a copper electrode [16]. This separation system allowed to profile different dextran compositions. While only the lowest molecular weight dextran gave a narrow band of closely spaced peaks, the larger dextrans exhibited a much broader manifold of peaks (Figure 5.8). From these results it was assumed that only the smaller dextran is composed of a well defined group of polysaccharides whereas the larger ones present a broad mixture of polysaccharides.

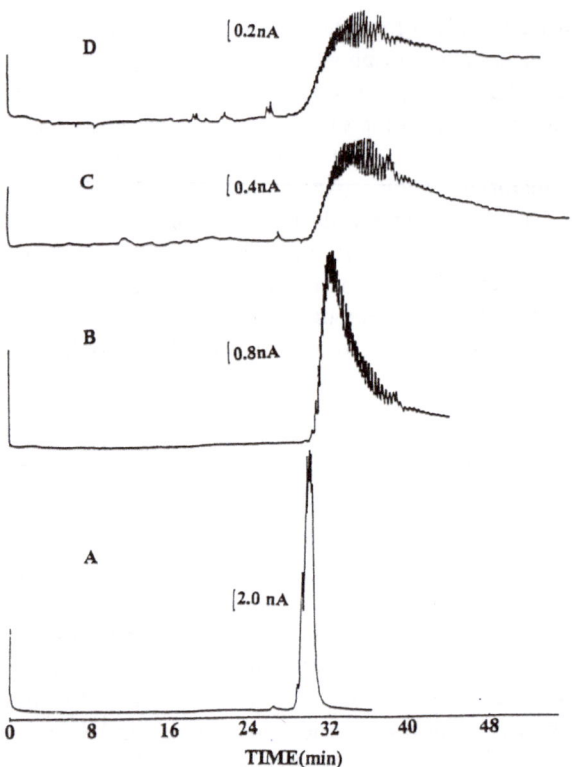

Figure 5.8 Electropherograms of dextrans with average molecular weights of (A) 1500, (B) 6000, (C) 9300, (D) 18300, each at 3 mg/ml concentration. Separation conditions: BGE, 200 mM NaOH with 10 mM CTAB; capillary, 80 cm x 25 µm ID; voltage, 8 kV; amperometric detection, working electrode: 127 µm Cu disk at +0.60 V *vs.* Ag/AgCl (from [16], with permission).

Pectin, a polysaccharide found in plant cell walls, has a backbone consisting mainly of GlcA, Rha, Ara and Gal. Typically 200–1000 GalA units, which are esterified to varying degrees, are linked through α-1,4-glycosidic bonds. Due to its charge and UV absorbing properties imparted by the carboxylate groups pectins can be subjected to electrophoretic separations as demonstrated by Zhong et al. [47]. The migration of the pectin was found to be a function of the degree of esterification, with a linear relationship between the charge of the pectin and its electrophoretic mobility. The CE method was used to quantify pectin in aqueous solutions in the concentration range of 0.5–5 mg/ml at 192 nm.

The electrophoretic migration of neutral and highly charged polysaccharides in open tubular CE, exemplified by chemically modified cellulose, such as hydroxypropylmethylcellulose (HPMC) or carboxymethylcellulose (CMC), and by heparins can be regulated by secondary thermodynamic equilibria using buffer additives [48]. Electrophoretic mobility of modified cellulose, which is rather low under 'normal' borate complexation conditions due to the modification of the majority of hydroxyl groups, can be increased by addition of hydrophobic and charged compounds to the running buffer. For instance SDS forms a monolayer on HMPC and CMC resulting in very fast migration of these compounds but also in low resolution. Therefore competing agents such as hydrophobic compounds or ion-pairing reagents need to be added in order to enhance the separation selectivity. The electrophoretic mobility of highly charged polysaccharides like heparin could be controlled by addition of ion-pairing reagents to the BGE. It decreased as the charge and concentration of cationic compounds was increased.

Nowadays a number of research projects are focused on the use of plant starch as industrial raw material in order to reduce dependence on non-renewable resources. A new approach for starch analysis uses iodine complexation to impart a charge to the starch components and to permit their detection in CE [49]. Amylopectin and amylose were resolved in less than 10 min. using an iodine containing BGE (Figure 5.9). The starch iodine complex consists of a helix of sugar residues surrounding a linear I_5^- core. Iodine binding exhibits strong chain-length dependence in both complexation and optical properties and is temperature dependent. Although the binding constant of iodine for amylose is several orders of magnitude higher than that of amylopectin both polymers bind approximately 20% of their molecular mass in iodine, a sufficiently high iodine concentration and a temperature of 20°C provided. The iodine complexation was found to be applicable to the CE separation of carbohydrates ranging from maltooligosaccharides of dp > 40 up to amylopectin with molecular masses in the tens of millions.

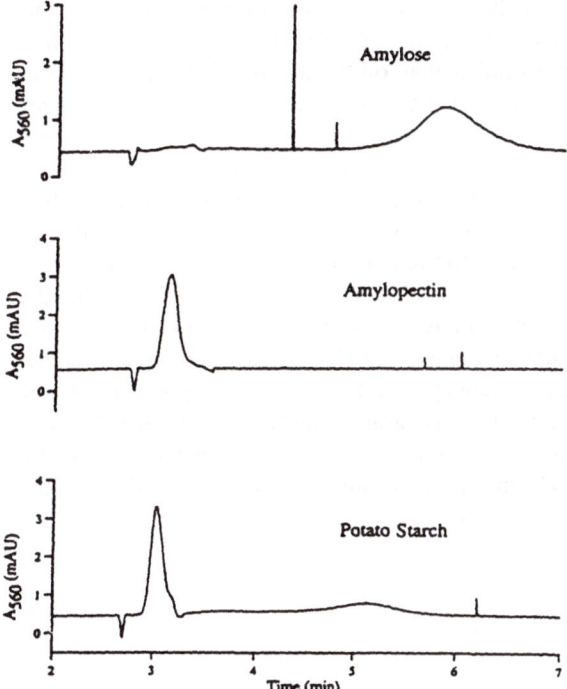

Figure 5.9 Electropherogram of amylose, amylopectin and potato starch. Separation conditions: BGE, 20 mM citrate-phosphate, pH 6.0, 0.1 mg/ml KI, 0.1 mg/ml I_2; capillary, 36/28 cm x 50 μm ID; voltage, 20 kV; temperature, 25°C; injection, 5s hydrodynamic; detection, 569 nm; sample concentration, 1 mg/ml (from [49], with permission).

5.4 Glycopeptides and Glycoproteins

5.4.1 Glycoforms

Protein glycosylation can occur at one or more positions in the amino acid sequence. Additionally, the glycans at a single position may be heterogeneous or may be missing at some positions. This leads to the formation of populations of glycosylated variants of a single protein called 'glycoforms' [50]. The formation of these glycoforms is affected by several factors such as the environment in which the protein is glycosylated, the manufacturing process and the isolation procedure. Therefore, all these factors affect the glycoform population and thus the functioning of the protein [51, 52]. However, as long as the production conditions are kept constant, the relative proportions of the glycoforms were found to be reproducible. Consequently, high-resolution separation methods allowing for the monitoring of such glycoform populations are in demand, in order to control biotechnologically manufactured recombinant glycoproteins as well as to investigate their biological functions. The huge number of applications cited in this chapter reflects the important role CE plays in the area of glycoprotein analysis.

The fact that for each glycoprotein individual separation systems, varying in pH and composition of the BGE and in the nature of the capillary are described, demonstrates the complexity in the analysis of these species. The resolution of glycoprotein glycoforms relies on the differences in the glycosylation pattern rather than on the nature of the protein backbone. Table 5.3 gives a comprehensive overview of the glycoproteins analyzed by CE and the separation conditions applied.

Early work on glycoprotein analysis focused on the separation of sialylated glycoforms. Kilár and Hjertén [53] were the first to report a fundamental study on the separation of human transferrin glycoforms by CZE and CIEF. The separation was based on differences in the number of sialic acid residues and resulted in the resolution of five components corresponding to the di-, tri-, tetra-, penta- and hexasialotransferrin. Using a polyacrylamide coated capillary for EOF suppression resulted in better resolution and sharper focusing of closely related glycoforms. The authors also demonstrated the gradual removal of sialic acids by neuraminidase treatment resulting in changes of the relative proportions of the different transferrin isoforms by time in the electropherogram (Figure 5.10). Charge heterogeneity due to variations in sialylation was also monitored in recombinant human growth hormone (rhGH), T4 receptor protein (rCD4) and tissue plasminogen activator (rtPA) using a coated fused silica capillary and a low pH BGE [54].

Table 5.3 Analytical conditions for CE analysis of glycoprotein glycoforms

Glycoprotein	Separation mode	Capillary	BGE	Detection (UV in nm)	References
Human transferrin	CZE	polyacrylamide coated	18 mM Tris – 18 mM boric acid – 0.3 mM EDTA, pH 8.4	280 nm	[53]
	CIEF	polyacrylamide coated	Ampholyte (2% Biolyte 5/7)	280 nm	
T4 receptor protein	CZE	coated with a polymer	phosphate, pH 4.5 or 5.5	200 nm	[54]
Recombinant human growth hormone	CZE	coated with a polymer	phosphate, pH 6.5	200 nm	[54]
Recombinant human erythro-poietin	CZE	untreated	100 mM acetate-phosphate, pH 4.0, capillary was preincubated for 10 hours	214 nm	[55]
Recombinant human erythro-poietin	CZE	untreated	10 mM tricine – 10 mM NaCl – 2.5 mM DAB – 7 M urea	214 nm	[56]
Recombinant human erythro-poietin	CZE	eCAP Amine[a]	200 mM phosphate – 1 mM nickel chloride, pH 4.0	200 nm	[57]
Ribonuclease A and B	CZE	untreated	20 mM CAPS, pH 11.0	200 nm	[58]
Ribonuclease A and B	CZE	untreated	20 mM phosphate – 5 mM tetra-borate – 50 mM SDS, pH 7.2	200 nm	[59]
Ovalbumin	CZE	untreated	25 mM tetraborate – 100 mM boric acid – 1mM DAB, pH 8.5	200 nm	[61]
Ovalbumin	CZE	untreated	25 mM tetraborate titrated with 100 mM boric acid to pH 8.4 – 0.3 mM HxBr, HxCl or DcBr	200 nm	[62]
Ovalbumin	CZE	untreated	25 mM or 100 mM borate, pH 9.0 - varying concentrations of monoamines, polyamines and SDS as EOF modifier	200 nm	[63]
Human chorionic gonadotropin	CZE	untreated	25 mM borate – 5 mM diamino-propane, pH 8.8	200 nm	[64]
Human chorionic gonadotropin	CZE	untreated	200 mM borate – 12.5 mM diaminopentane, pH 9.0	200 nm	[65]
Recombinant tissue plasmi-nogen activator	CIEF	coated with co-valently bonded linear polymer	2% ampholyte (pH 6-8) – 2% CHAPS – 6M urea	280 nm	[66]
	CZE	coated with covalently bonded linear polymer	0.1M ammonium phosphate, pH 4.6 – 0.01% TritonX-100	200 nm	

Table 5.3 continued

Glycoprotein	Separation mode	Capillary	BGE	Detection (UV in nm)	References
Recombinant tissue plasminogen activator	CZE	polyacrylamide or polyvinylalcohol coated	200 mM n-aminocarboxylic acid buffered with 200 mM acetic acid, pH 4.6 or 5.1	214 nm / 280 nm	[67]
	CIEF	eCAP neutral capillaries[a]	4% ampholyte (pH 3–10) – 36% urea in cIEF 3–10 Kit polymer solution[a]	214 nm	
	SDS-CGE	eCAP SDS coated[a]	SDS 14–200 gel buffer[a]		
Human serum transferrin	CZE	polyacrylamide coated	18 mM Tris – 18 mM borate – 0.03 mM EDTA, pH 8.4, 0 – 8 M urea	280 nm	[71]
Human serum transferrin	CZE	untreated	25 mM tetraborate titrated with 100 mM boric acid to pH 8.3 – 1 mM DAB, 0.75 mM HxBr or 1–2 mM DcBr	200 nm	[72]
Human serum transferrin	CGE	DB-17 coated	100 mM borate, pH 8.5 – 0.5% HEC	200 nm	[73]
Recombinant human interferon-γ	MECC	untreated	400 mM borate – 100 mM SDS, pH 8.5	200 nm	[75]
Recombinant interferon-ω	MECC	polyacrylamide coated	150 mM boric acid – 50 mM SDS, pH 9.7	200 nm	[52]
Monoclonal immuno globulin G	CZE	untreated	20 mM CHES, pH 9.5	200 nm	[76]
Monoclonal antibody	CZE	untreated	150 mM tetraborate, 9.4	200 nm	[77]
α_1-Acid glycoprotein	MECC	untreated	50 mM phosphate – 10 mM tetraborate – 40 mM SDS, pH 9.0	200 nm	[78]
Proteinase	CZE	untreated	100 mM acetate phosphate, pH 3.2	200 nm	[79]
Recombinant human bone morphogenetic protein 2	CZE	coated[b]	100 mM phosphate, pH 2.5	200 nm	[80]
Recombinant human bone morphogenetic protein 2	CZE	polyacrylamide coated	50 mM β-alanine, pH 3.5 (with acetic acid)	200 nm ESMS	[90]
Recombinant human factor VIIa	CZE	untreated	100 mM phosphate – 25 mM DAB, pH 8.0	214 nm	[81]

Table 5.3 continued

Glycoprotein	Separation mode	Capillary	BGE	Detection (UV in nm)	References
Recombinant human granulo-cyte-colony stimulating factor	CZE	untreated	50 mM phosphate - 50 mM borate, pH 8.0 – 100 mM borate, pH 9.0	214 nm	[82]
Surface glyco-proteins of ovine lentiviruses	CZE	untreated	20 mM phosphate, pH 9.0	200/214 nm	[83]
Mucin	CGE	untreated	100 mM CAPS, pH 8.8 – 0.5% PEG	200 nm	[84]
Horseradish Peroxidase	CZE	Polybrene coated	2 M formic acid	200 nm ESMS	[89]
Desmodus sali-vary plasmino-gen activator	CZE	BSA treated	100 mM phosphate, pH 3.0	200 nm MALDI	[92]
Hirudin	CZE	untreated	90 mM boric acid, 15 mM tetra-borate – 0.2 mM DAB, pH 8.3	200 nm	[168]

CAPS = 3-(cyclohexylamino)-1-propanesulfonic acid, CHAPS = 3-((3-Cholamidopropyl)-dimethyl-ammonio)propanesulfonic acid, CHES = 2-(Cyclohexylamino)-ethanesulfonic acid, PEG = polyethy-lene glycol, DAB = 1,4 diaminobutane, HEC = hydroxyethyl cellulose, HxBr = hexamethonium bro-mide, HxCl = hexamethoniumchloride, DcBr = decamethonium bromide, BSA = borine serum albu-min.

[a] Product of Beckman Instruments, Fullerton, CA, USA. [b] Precoated capillary from Bio Rad.

5.4.1.1 Erythropoietin

Similar studies were performed on recombinant human erythropoietin (rhEPO), a 35'000 Da glycoprotein hormone which is primarily produced in the kidney of adult mammals and which promotes development of mature red blood cells [55]. The carbohydrate part of the protein accounting for up to 40% of its molecular mass, comprises three N- and one O-gly-cosylation sites carrying glycan chains extensively differing in their degree of sialylation. Resolution of the glycoforms was optimized by systematically changing pH, the nature of the BGE and the organic modifier. The best resolution was obtained using a 100 mM aceta-te-phosphate buffer (pH 4.0) and a 10 hours preincubation of the capillary with the separa-tion buffer. A more elaborate study on the separation of rhEPO glycoforms was presented by Watson and Yao [56]. They attempted to improve the resolution by reduction of the EOF through addition of polymeric modifiers (e.g. polyethyleneglycol), of organic solvents and of small basic amines such as 1,4-diaminobutane or 1,5-diaminopentane. This approach is often applied in glycoprotein analysis to increase the resolution of closely related species. Balancing the EOF against the electrophoretic migration enhances differences in the charge-to-mass ratio of the solutes and thus the resolution of the glycoforms. While in a tricine/

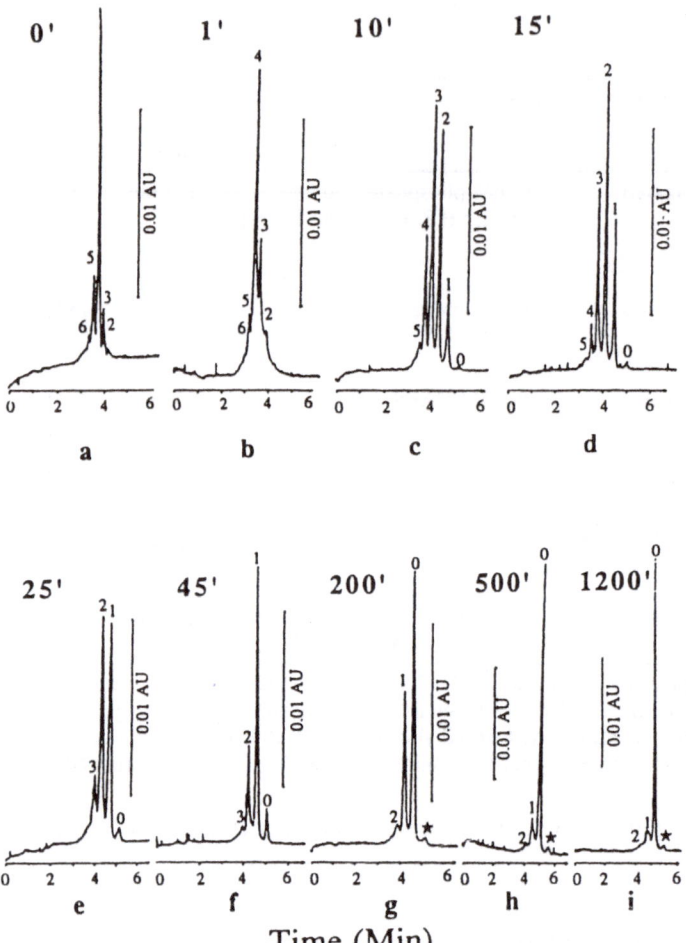

Figure 5.10 CZE of iron-free transferrin following incubation with neuraminidase. The samples were taken after various incubation times: (a) 0, (b) 1, (c) 10, (d) 15, (e) 25, (f) 45, (g) 200, (h) 500 and (i) 1200 min. Separation conditions: 18 mM Tris-boric acid, 0.3 mM EDTA, pH 8.4; capillary, 18.5 cm x 100 µm ID; voltage, 8 kV; UV detection, 280 nm (from [53], with permission).

NaCl buffer the EPO glycoforms migrated in only one single peak, the addition of 1,4-di-aminobutane resulted in the resolution of four distinct peaks (Figure 5.11 A+B). Addition of high concentrations of urea improved the resolution further due to reduced adsorption of the glycoprotein on the capillary wall (Figure 5.11 C). The glycoforms were separated according to an increased number of sialic acids. This was confirmed by comparing the separation patterns obtained with CE and gel IEF and by spiking with individual glycoforms isolated by preparative IEF. Additionally, it was demonstrated that after incubation with neuraminidase the peak of the slowest migrating species decreased first, indicating that this species carries the highest number of sialic acid residues.

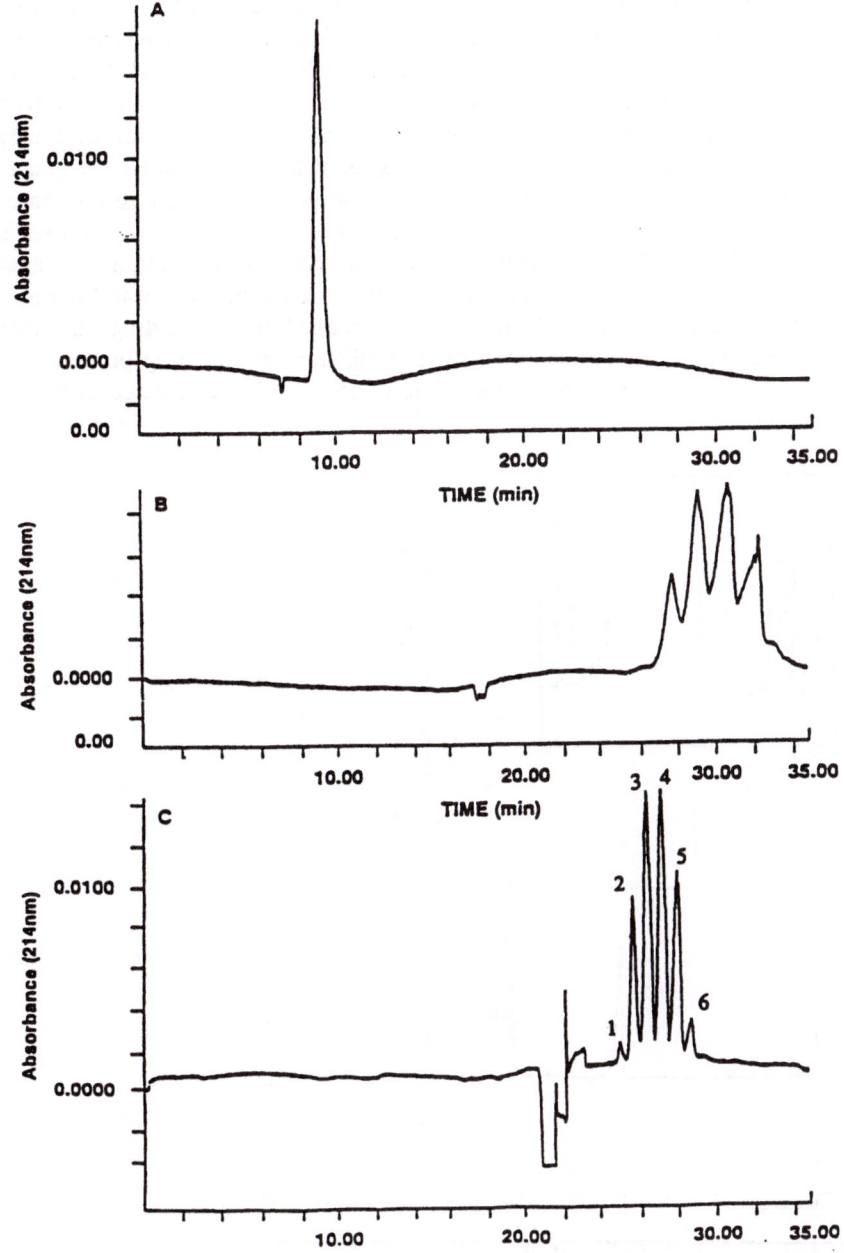

Figure 5.11 Capillary electrophoretic separation of r-HuEPO at 1 mg/ml. Separation conditions: BGE, (A) 10 mM tricine – 10 mM NaCl, pH 6.2, (B) 10 mM tricine – 10 mM NaCl – 2.5 mM 1,4-diaminobutane, pH 6.2, (C) 10 mM tricine – 10 mM NaCl – 2.5 mM 1,4-diaminobutane – 7 M urea, pH 6.2; capillary, 50 cm x 75 µm ID; voltage, 10 kV; UV detection, 214 nm (from [56], with permission).

Very recently, a CE method for the analysis of recombinant human EPO in final drug preparations was presented [57]. All products investigated were formulated with large amounts of human serum albumin (HSA) which is used to stabilize the rhEPO. Since both proteins show very similar physical characteristics in solution they could not be resolved under CE conditions developed for the separation of bulk rhEPO. But addition of 1 mM nickel chloride to the separation electrolyte, which consists of 200 mM sodium phosphate, pH 4.0, allowed for both, the complete separation of the two proteins and the resolution of the rhEPO glycoform population (Figure 5.12). Quantitation was found to be linear over a concentration range of 0.03–1.92 mg/ml, with a limit of quantitation of 0.03 mg/ml. An analysis of the products from two manufacturers exhibit only little qualitative lot-to-lot variations in the rhEPO glycoform patterns, indicating good reproducibility of the drug substance manufacturing process. However, the quantitative determination of two drug products revealed a significant variability with lot-to-lot variations of 25 and 33%, calculated as rhEPO area counts per 1000 IU.

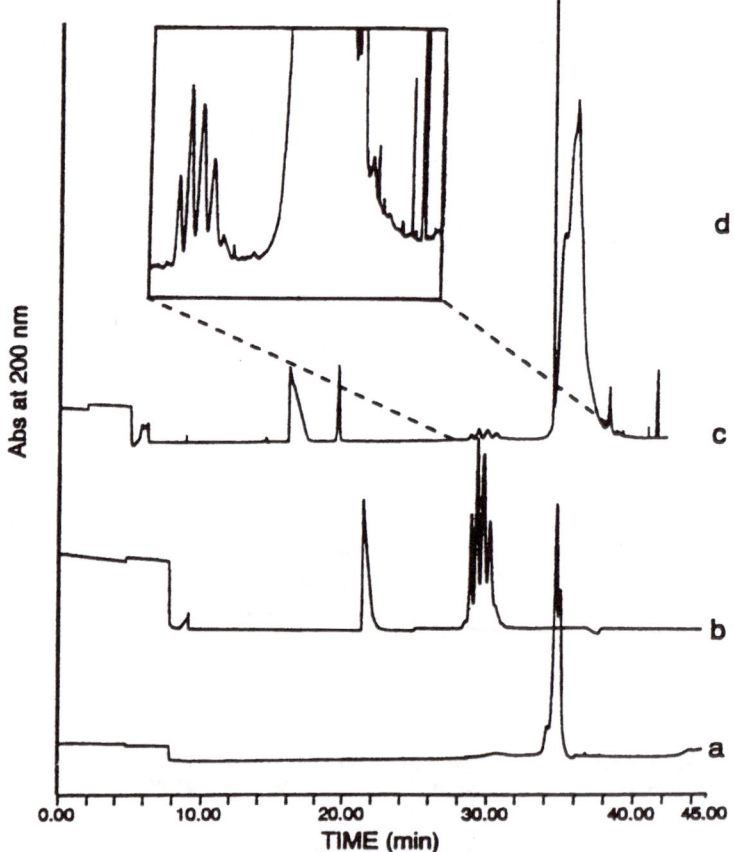

Figure 5.12 Electropherograms of (a) HSA, (b) bulk EPO and (c) rhEPO formulation. Inset: expanded view of the region between 30 and 45 min. in electropherogram (c). Separation conditions: BGE, 200 mM sodium phosphate, pH 4.0, 1 mM nickel chloride; capillary, 47/40 cm x 50 μm ID; voltage, 8 kV; temperature, 20°C; injection, 8 sec with 0.5 p.s.i.; UV detection, 200 nm (from [57], with permission).

5.4.1.2 Ribonuclease

Grossmann et al. [58] described the separation of a mixture of ribonuclease (RNase) A, B1 and B2. Whilst RNase A is not glycosylated, RNase B1 and B2 differ in the extent of glycosylation at the asparagine (asn) 34. The peaks were identified by separating the three components on a concanavalin A affinity column and then analyzing each component individually by CE. Using a phosphate buffer (pH 7.2) containing SDS and tetraborate, Rudd et al. [59] succeeded in resolving the different glycoforms associated with RNase B. This glycoprotein contains high-mannose type oligosaccharides with 5–9 mannose residues (Man5–Man9) attached to its single N-glycosylation site. The relative proportions of the various glycoforms determined by this method correlated well with those determined by other methods like mass spectrometry and HPAEC as indicated in Table 5.4. The authors assumed the separation to occur mainly based on the formation of borate complexes with the outer mannose residues. High-mannose type linkage isomers were not separated in this buffer since they complex with the borate ion to approximately the same extent [60]. To further substantiate the presence of the different glycoforms, the time dependence of the α-1,2-mannosidase digestion was monitored, cleaving mannose residues from the non-reducing end of the glycan. After 25 hours of incubation the glycoform population carrying Man5–Man9 structures was reduced to one single population carrying Man5 [59].

Table 5.4   The relative proportions of the glycoforms of ribonuclease B determined by (a) mass spectrometry, (b) Bio-Gel P-4 gel filtration, (c) HPAEC with radioactive detection, and (d) CE [59]

Glycoforms	(a) MS [%][a]	(b) P-4 [%][a]	(c) HPAEC [%][a]	(d) CE [%][a]
Man5	47	49	49	48
Man6	21	19	19	20
Man7	11	11	10	11
Man8	16	17	16	17
Man9	5	4	6	4

[a] Numbers are relative area percent.

5.4.1.3 Ovalbumin

Ovalbumin, a glycoprotein isolated from hen egg white, was also analyzed by CE for its glycoform components [61]. Ovalbumin contains one N-glycosylation site which accommodates at least nine different oligosaccharide structures of the high mannose and the hybrid type. Furthermore, ovalbumin contains two potential phosphorylation sites. Using a 100 mM borate buffer (pH 8.5) containing 1 mM 1,4-diaminobutane for low EOF conditions, the ovalbumin glycoforms were well separated. Dephosphorylation studies by treating the glycoprotein with alkaline and acidic phosphatase revealed that the phosphorylation did not affect the resolution but caused retardation of the migration time. From these results it was suggested that all glycoforms are phosphorylated to the same degree and microheterogeneity relies solely on differences in the carbohydrate structures. In a more recent study Oda et al. [62] investigated the separation of the ovalbumin glycoforms in the presence of α,ω-bis quaternary ammonium alkanes as buffer additives. The alkyl chain length and the cation group of the α,ω-bis quaternary ammonium alkanes strongly influenced the analysis time and the resolution. Under identical separation conditions Decamethoniumbromide $C_{10}MetBr$ was shown to yield shorter analysis time and somewhat better resolution than Hexamethonium-bromide C_6MetBr (Figure 5.13). The fact that all additives investigated, including 1,4-di-aminobutane [61], repressed the EOF similarly but the quaternary ammonium alkanes allo-wed resolution of the ovalbumin glycoforms in half the time compared to diaminobutane suggests that other mechanisms than EOF repression may be involved such as additive-pro-tein interactions, protein-wall interactions etc.

The effect of polyamine, diamine and monoamine buffer additives on the resolution of ovalbumin glycoforms was also investigated [63]. It was demonstrated that changes in selec-tivity are the consequence of many factors including molecular weight, concentration and molecular structure of the modifier but not of the EOF itself. The selectivity factor increased with an increase in molecular weight, number of methyl or methylene groups, the ratio of methyl or methylene groups to amino groups and the total length of the modifier chain. The EOF was shown to increase linearly in the same direction. Optimum resolution of the oval-bumin glycoforms was achieved using a 25 mM borate buffer (pH 9.0) containing 0.87 mM spermidine or 0.14 mM spermine.

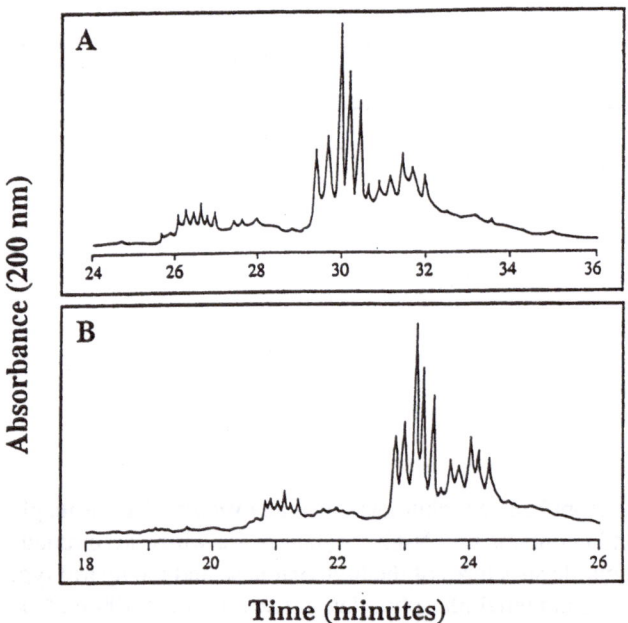

Figure 5.13 A comparison of the effect of a 300 μM concentration of each of the bis-quaternary ammonium additives on the resolution of ovalbumin. (A) 300 μM C_6MetBr, (B) 300 μM $C_{10}MetBr$. Separation conditions: BGE, 100 mM borate, pH 8.4; capillary, 87/80 cm x 50 μm ID; voltage, 25 kV; temperature, 28°C; injection, 3 sec, 0.5 p.s.i.; UV detection 200 nm. Sample concentration, 1 mg/ml (from [62], with permission).

5.4.1.4 Chorionic Gonadotropin

The α,ω-bis quaternary ammonium alkanes used by Oda et al. [62] for the analysis of oval-bumin were also applied to the separation of the glycoforms of human chorionic gonadotro-pin (hCG), a heteromeric glycoprotein consisting of a α- and a β-subunit [64]. Both subunits contain two N-glycosylated asparagines and the β-subunit an additional O-glycosylated serine. Using a 25 mM borate buffer (pH 8.8) and 1 mM C_6MetBr seven of eight glycoforms were resolved. Exchanging the C_6MetBr with diaminopropane allowed almost baseline reso-lution of all eight expected glycoforms in less than 50 min. In the absence of diaminopro-pane only one single broad peak was observed. When analyzed individually the α-subunit separated into four and the β-subunit into seven distinct peaks. The potential clinical appli-cation of this method was demonstrated by analyzing hCG derived from four different sour-ces. While the number of different isoforms was constant among the samples the relative concentration of the individual isoforms varied. Also using a borate buffer system, Laidler et al. succeeded in resolving six major and eight minor hCG isoforms [65]. Following the treatment with neuraminidase, the series of peaks observed with the untreated material was replaced by a set of faster migrating peaks. This observation is consistent with a removal of sialic acids correlated with a reduction in the number of negative charges in the hCG iso-forms. Since the degree of sialylation of hCG is correlated with its biological activity this CE assay may be used to assess the hormone potency.

5.4.1.5 Tissue Plasminogen Activator

A glycoprotein under extensive investigation by CZE and CIEF is the recombinant tissue plasminogen activator (rtPA), a fibrin specific protein that has been approved for treatment of myocardial infractions. The two main glycosylation variants of rtPA are type I which carries three N-glycosylation sites at asn-117, -184 and -448 and type II which is glycosyl-ated at asn-117 and -448. Asn-117 is associated with high mannose type oligosaccharides only, while for asn-448 complex type bi-, tri- and tetraantennary oligosaccharides are found. As expected both, the CZE and CIEF separation of type I and II rtPA exhibit different elec-trophoretic patterns as illustrated in Figure 5.14 [66]. The microheterogeneity of this glyco-protein was manifested by the partial resolution of almost 15 glycoforms of a protein with only four potential glycosylation sites. The comparison of the CZE- and the CIEF-profile, respectively, of the intact rtPA with that of a desialylated rtPA revealed a much simpler profile for the latter one. This indicates that the microheterogeneity observed with rtPA is mainly the result of different degrees of sialylation.

CZE and CIEF of rtPA glycoforms was also the subject of a very recent study [67]. The authors addressed several issues important to the validation of methods used in the charac-terization of recombinant proteins, including the protein recovery and the migration time reproducibility. To determine the protein recovery CZE and CIEF eluates were analyzed by means of an ELISA test. The use of an ε-aminocaproic acid based BGE and a polyvinylal-cohol (PVA) coated capillary allowed fast separation of the rtPA glycoforms. Addition of Tween 80 resulted in an almost 100% recovery and a t_M-reproducibility of $\leq 0.2\%$ RSD. The use of neutrally coated capillaries to minimize the EOF, a part of it filled with a polymer solution, enhanced the reproducibilty of the present CIEF method.

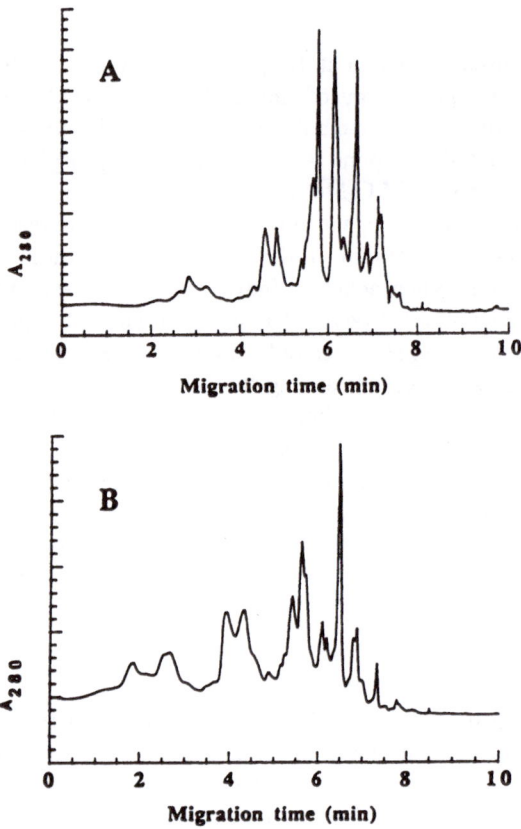

Figure 5.14 IEF patterns of (A) Type I rtPA and (B) Type II rtPA: Separation conditions: BGE, 2%
ampholyte (pH 6–8), 2% CHAPS – 6 M urea; capillary, 14 cm x 25 μm ID, Bio-Rad
coated; voltage, 12 kV for focusing, 8 kV for mobilization; UV detection 280 nm (from
[66], with permission).

The suitability of different ampholytes for CIEF of rtPA was tested and Ampholine 3.5–
10 was found to give the sharpest peaks and the best resolution as illustrated in Figure 5.15
[67]. Chen et al. [68] who evaluated four commercial ampholytes (Ampholine, Pharmalyte,
Bio-Lyte and Servalyte) regarding the resolution of rtPA glycoforms ended up with almost
the same result. Using Ampholine 3.5–10, rtPA was resolved in several major and minor
bands with different patterns for type I and type II [67]. Addition of urea to the CIEF sample
was necessary to enhance protein solubilization and to provide full recovery. The authors
stated that due to the high t_M-reproducibility and protein recovery CZE and CIEF are suitable for routine analysis of glycoform heterogeneity. The feasibility of using denaturing
SDS-CE with coated capillaries and polymer additives for sieving to estimate the molecular
weight of rtPA variants was demonstrated in the same report [67]. It was assumed that this
method, performed using a commercial kit, could be an alternative to the time consuming
SDS-polyacrylamide gel electrophoresis (PAGE).

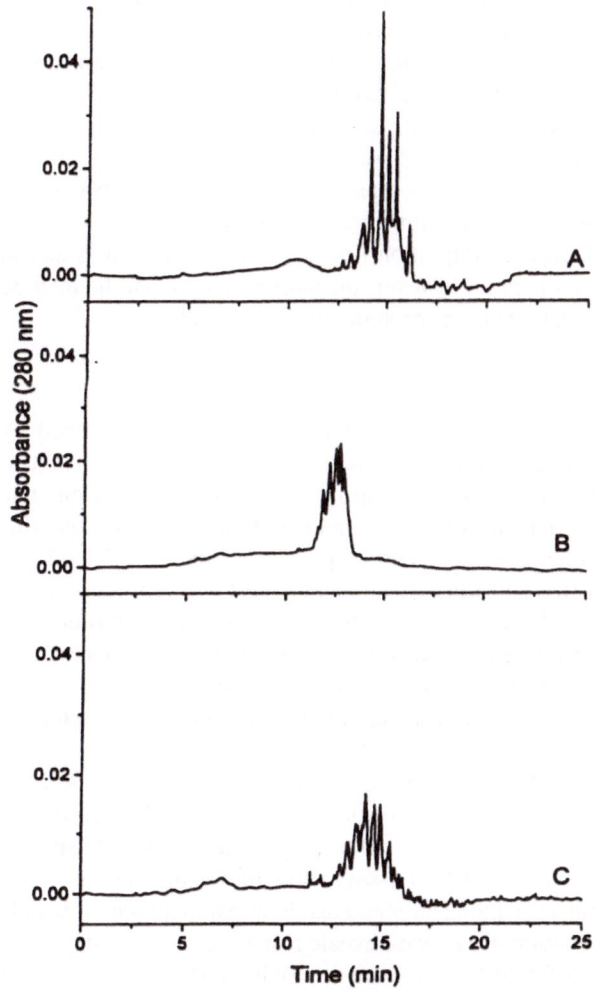

Figure 5.15 Comparison of ampholytes for the separation of rtPA. (A) Ampholine 3.5–10, (B) Pharmalyte 3–10, (C) Servalyte 3–10. Separation conditions: anolyte, 91 mM phosphoric acid in gel buffer; catholyte, 10 mM NaOH; capillary, 27/20 cm x 50 µm ID, neutral coated; voltage; 500 V/cm for focusing, 500 V/cm with low pressure for mobilization; UV detection, 280 nm. Sample: 0.32 mg/ml rtPA, 4.8 M urea and 1.3% ampholyte (from [67], with permission).

Moorhouse et al. [69] developed a CIEF method for rtPA analysis based on an one-step approach where focusing and mobilization occur simultaneously. The focusing takes place between detector and anode and the EOF sweeps the separated glycoforms past the detector. Inclusion of hydroxypropylmethylcellulose and urea in the separation matrix were found to be the key factors for achieving maximum resolution while maintaining protein solubility during the focusing step. The charge heterogeneity exhibited by rtPA could be detected as a series of approximately 10 peaks. The increasing migration times corresponded to the increasing acidic pI's of the individual charged species. In a further study this rtPA-method was validated [70]. The method showed an acceptable recovery of >100% , determined by ELISA, and good sensitivity with 25 ng of glycoprotein still being resolved into constituent peaks. The recovery proved to be linear over a range of 50–1000 µg/ml protein. Intra-assay precision for six sequential injections was 3% for both migration time and peak area. Regarding the uncoated separation capillaries the method was found to be not rugged. Whilst CIEF profiles generated with one capillary lot were comparable, they differed between lots in terms of resolution of peak shoulders and migration time of more acidic species. Additionally, capillary aging resulted in changes in migration time on a day-to-day basis and in inconsistent resolution of the more minor peaks. However, the authors stated that the method has the potential to be used in commercial release of protein pharmaceuticals.

5.4.1.6 Transferrin

The suitability of CE to monitor conformation changes of proteins was demonstrated by Kilár and Hjertén [71] by measuring the unfolding of human serum transferrin in urea. Unfolding of iron-free transferrin is a two-step process going through a stable transition state which is called the unfolding state. In a denaturing environment containing 3–6 M urea, this unfolding state coexists with the unfolded state for a certain period of time. The folded and the unfolded conformation of the iron-free isoforms are characterized by different migration times. By plotting the migration time vs. the urea concentration in the separation buffer a denaturation curve of the transferrin forms was constructed (Figure 5.16). The iron-saturated transferrin exhibits a more compact conformation and was found be more resistant to urea denaturation. Unlike iron-free transferrin with the iron-saturated form no unfolding intermediates were observed.

Similar to the first study of Kilár and Hjertén on human transferrin [53], Oda and Landers [72] investigated the microheterogeneity of transferrin associated with the varying sialylation of its glycoforms. They focused on the analysis of transferrin from different species as a means of evaluating the potential for a CE-based assay, allowing for the determination of carbohydrate deficient transferrin (disialo, monosialo and asialo transferrin) which is only observed under certain pathological conditions. It was revealed that for successful separation of the transferrin isoforms from different species such as bovine and human, different separation conditions were required. While bovine transferrin isoforms were resolved in a borate buffer containing 1,4-diaminobutane, the same conditions were found to be inadequate for the resolution of the sialo-forms from human transferrin. The authors stated that the optimal conditions for CE separations are not only glycoprotein specific but even species specific for the same glycoprotein. In a subsequent paper the authors further demonstrated the suitability of a CE based assay as a diagnostic tool for clinical syndromes correlated with the presence of carbohydrate deficient transferrin in human serum [73]. Using a DB-17-coated capillary, allowing for reduced EOF separation conditions, and a sieving buffer containing hydroxyethyl cellulose, separation of all transferrin sialoforms was achieved within eight minutes. Comparing the transferrin sialoform pattern of normal patients and patients suffering from carbohydrate defficient glycoprotein syndrome (CDGS) significant differences were revealed (Figure 5.17). Separation of transferrin isoforms under non-denaturing conditions and basic pH allowed for the investigation of plasma samples from a population

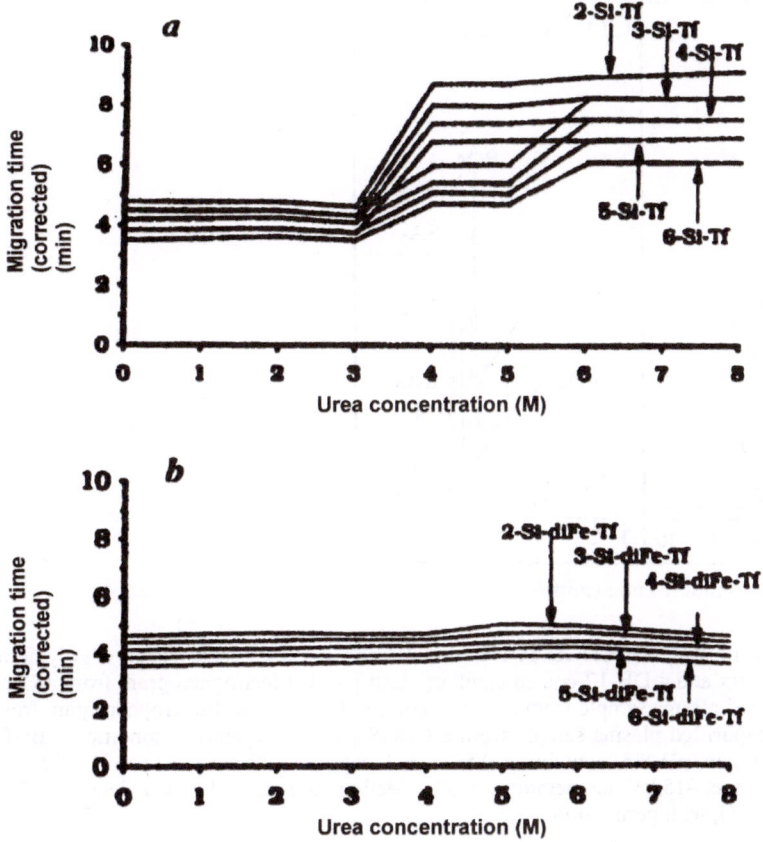

Figure 5.16 Denaturation curves of the human serum (a) iron-free and (b) diferric (dife) transferrin isoforms between 0 and 8 M urea. Corrected migration times for the 2-sialo-, 3-sialo-, 4-sialo-, 5-sialo- and 6-sialo-transferrin forms (marked as 2-Si-Tf, 3-Si-Tf, 4-Si-Tf, 5-Si-Tf, 6-Si-Tf or 2-Si-diFe-Tf, 3-Si-diFe-Tf, 4-Si-diFe-Tf, 5-Si-diFe-Tf, 6-Si-diFe-Tf, respectively) are plotted *vs.* the urea concentration (from [71], with permission).

consuming varying amounts of alcohol at different intervals [74]. A cut-off value of 3% carbohydrate deficient transferrin resulted in a clinical sensitivity of 88% in a population consuming at least 70 g/day alcohol for a minimum of two weeks. In a population consuming less than 70 g/day the sensitivity decreased significantly. These results confirmed carbohydrate deficient transferrin as a specific marker for significant and chronic use of alcohol.

5.4.1.7 Interferon

A MECC based separation system was recently applied to resolve glycoform populations of human interferon-γ produced by Chinese hamster ovary (CHO) cells [75]. Human interferon-γ exhibits variable site-occupancy at asn-25 and asn-29 with the microheterogeneity

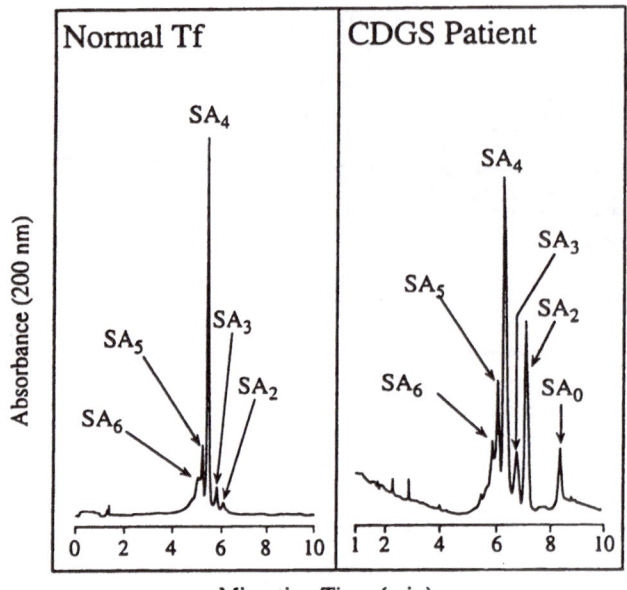

Figure 5.17 Comparing normal and CDGS Tf sialoforms using rapid CE analysis in a polymeric sieving matrix and a DB-17-coated capillary. Left panel: Electropherogram from an immunopurified plasma sample from a normal control. Right panel: Electropherogram from an immunopurified plasma sample from a CDGS patient. Separation conditions: BGE, 100 mM borate, pH 8.5, containing 0.5% HEC, capillary, 27 cm x 50 µm ID, DB-17-coated; voltage, –15 kV, temperature, 20°C; injection, 20 sec, 3.5 kV; UV detection, 200 nm (from [73], with permission).

being a result of the presence or absence of terminal sialic acid and core fucose. Using a high ionic-strength borate/SDS buffer allowed for the resolution of more than 30 peaks which were classified into three groups PG1, PG2 and PG3, according to their migration time. In case of the SDS containing separation buffer the migration time of the different glycoforms was inversely related to the amount of carbohydrate associated with the protein. Progressive removal of N-glycans by peptide-N-glycosidase F (PNGase F) digestion of the Interferon-γ glycoforms suggested that PG1 presents the doubly glycosylated and PG2 the singly glycosylated glycoforms whilst PG3 presents the proteins with no or little glycosylation. Digestion with neuraminidase and endoH showed most glycoforms to be associated with sialylated complex type and a minor proportion to be associated with high mannose type oligosaccharides. Kopp et al. [52] elucidated the robustness of interferon-ω-glycosylation expressed in CHO-cells cultivated under different fermentation conditions. During fermentation the most significant alterations resulted from various parameters such as the ammonia concentration in the production media, the cultivation mode and the process time. HPAEC and CE analyses revealed that increasing ammonia concentrations and a shortened process time resulted in a reduction of biantennary and triantennary oligosaccharide structures. Adherent cultivated cells showed the tendency towards synthesis of higher branched structures.

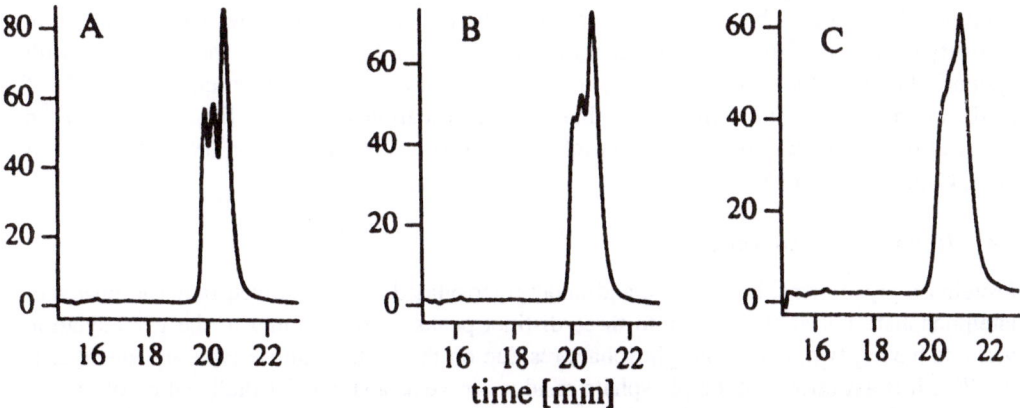

Figure 5.18 Separation profiles of an antibody (batch 1715) after storage in glass vials: (A) 3 months at 2 - 8 °C, (B) 3 months at room temperature, (C) 3 months at 37 °C. Separation conditions: BGE, 200 mM sodium tetraborate; capillary, 48 cm x 75 μm ID, voltage, 10 kV; temperature, 20 °C; injection, 2 sec, 50 mbar, UV detection, 200 nm (from [77], with permission).

5.4.1.8 Immunoglobulins

CE was also employed for the identification and determination of isoforms of monoclonal F(ab')$_2$ fragments obtained after pepsin proteolysis of immunoglobulin G (IgG) [76]. Variations in the pH of the BGE and the applied electrical field were used to modulate the selectivity of the isoform separation. Resolution of all five F(ab')$_2$ isoforms was achieved using an alkaline CHES buffer (pH 9.5) and a low field strength. Even isoforms with pI differences as little as 0.23 units were clearly resolved. The oligosaccharide mediated microheterogeneity of a monoclonal antibody was also studied by complex formation with borate [77]. Characteristic separation patterns were obtained with three distinct peaks exhibiting the same UV-spectra. Chemical hydrolysis of the glycans and digestion with PNGase F resulted in changed pattern for the antibody and showed the N-glycans to belong to the agalacto-, biantennary type, core-substituted with fucose. The separation method was validated for linearity and reproducibility of migration time and peak area. CE was found to be suitable to analyze batch-to-batch consistency in production and to test the stability of galenical formulations as demonstrated by analysis of antibody formulations stored for 3 months at 37°C (Figure 5.18). As assumed by the authors, the separation profile of the antibody changed distinctly due to degradation of the carbohydrate moieties.

5.4.1.9 α$_1$-Acid Glycoprotein

Using MECC with SDS, Rice et al. [78] developed a method to separate α$_1$-acid glycoprotein (AGP) from hyaluronan and a number of low-molecular mass components, such as uric acid, in human synovial fluid. The AGP in the synovial fluid was identified by coinjection with human AGP and by reaction with neuraminidase. The enzyme released the neuraminic acid from the oligosaccharide chains, resulting in a migration time shift of the AGP peak in the electropherogram. Additionally, a sample of the glycoprotein was isolated by micropreparative CE, separated by SDS-PAGE and incubated with AGP antibodies which resulted in a positive reaction of the isolated material. Using synovial fluid samples from a patient with systemic lupus erythematosis, a rheumatic disease characterized by high elevated levels of AGP, the authors could demonstrate that the identification and quantification of AGP in synovial fluids by CE may provide an useful indicator for the diagnosis in the early stages of various types of arthritis.

5.4.1.10 Various glycoproteins

Proteinase glycoforms, both native and underglycosylated, were resolved in a 100 mM acetate/phosphate buffer (pH 3.2) into three distinct peaks corresponding to the glycosylation variants having two, one or no phosphate residue on the high-mannose type glycans at asn-68 [79]. It is assumed that the phosphate residues serve as a signal for the location of proteinase A to the vacuole. Recombinant human bone morphogenetic protein 2 (rhBMP-2) is a disulfide linked homodimeric glycoprotein which induces bone formation *in vivo*. Using a coated capillary and a simple phosphate buffer the 15 glycoforms of rhBMP-2 with the formula $(rhBMP-2)_2-(GlcNAc)_4(Man_x,Man_y)$ could be separated into 9 peaks with 10 to 18 mannose units [80]. Since the differences between any adjacent peaks corresponds to one mannose, it was possible to resolve two 30'000 Da proteins differing by only 180 Da. This corresponds to a mass resolution of about 170. A combination of HPAEC and CE was used to analyze the glycoforms of human recombinant factor VIIa (hrFVIIa), a vitamin K-dependent glycoprotein that participates in the extrinsic pathway of blood coagulation. A diaminobutane containing BGE allowed separation of the hrFVIIa-glycoforms according to the differences in their sialic acid content [81]. HPAEC analysis of oligosaccharides released from neuraminidase treated hrFVIIa revealed that oligosaccharide heterogeneity occurs even without sialic acid. Separation of recombinant human granulocyte-colony-stimulating factor (rh-GCSF) which contains two O-linked glycans differing in the presence of one or two sialic acids was achieved in a pH 8.0 phosphate-borate buffer [82]. Schmerr and Goodwin [83] applied CE for the characterization of surface glycoproteins of ovine lentiviruses, a group of viruses infecting sheeps and goats. To delinate differences in the surface glycoproteins from several viral chains they were treated with β-N-acetylglycosaminidase, neuraminidase and various exo- and endoglycosidases. Subsequent CE-analysis generated characteristic electropherograms for each viral strain. Mucus glycoproteins or mucins which are the most important structural components of the mucus layer covering the intestinal mucosa, contain hundreds of oligosaccharide chains covalently attached through O-glycosidic bonds to their polypeptide backbone. Due to their large size, their multiple charges and their poor solubility mucins usually adsorb heavily to the capillary wall. But using a zwitterionic BGE and PEG as a polymeric additive mucin was resolved in various fractions allowing for the differentiation of cecal mucin glycoforms, derived from germ-free and gnotobiotic mice [84].

5.4.1.11 New approaches in CE glycoform analysis

To obtain accurate molecular weights for glycoproteins by SDS-PAGE has been difficult due to the lack of SDS binding of the carbohydrate moieties resulting in lower charge-to-mass ratios for the SDS-glycoprotein complexes and an overestimation of molecular

Table 5.5 Ferguson analysis of glycoproteins [85]

Protein	Carbohydrate content (%)	Literature[a] molecular weight	ProSort[b] molecular weight	Ferguson[c] molecular weight	SDS-PAGE[d] molecular weight
Avidin	19	17'000	23'000	16'000	19'000
RNAse B	10	15'000[e]	26'000	17'000	20'000
AGP	42	40'000	80'000	41'000	45'000
Fetuin	23	45'000	79'000	52'000	60'000
Glucose oxidase	17	78'000	83'000	70'000	67'000
Lactoferrin	8	77'000	90'000	78'000	76'000
Lacto-peroxidase	18	76'000	93'000	74'000	65'000
Butylcholinesterase	20	75'000	119'000	56'000	88'000

[a] Analytical ultracentrifugation
[b] Standard ProSort Protocol
[c] Ferguson protocol using four concentrations of ProSort SDS-Protein Analysis Reagent
[d] SDS-PAGE: 12% acrylamide monomer, 2.6% N,N'-methylenebisacrylamide cross-linker
[e] Determined by mass spectrometry

weights. To minimize these inaccuracies, Ferguson plots are employed which require the determination of relative mobilities of standard and unknown proteins at different gel concentrations. Werner et al [85] demonstrated a procedure which automatically generates all the data required for a Ferguson plot using a replaceable sieving matrix in the capillary format. Multiple gel concentrations were produced by simply diluting the sieving matrix with buffer. By analyzing RNAse B, α_1-acid glycoprotein, fetuin, glucoseoxidase, lactoferrin, lactoperoxidase and butylcholinesterase at four different gel concentrations and calculating the retardation coefficients, molecular weights were determined showing good correlation with literature values (Table 5.5). The time required to perform the whole Ferguson analysis in the automated capillary format was comparable to the time required to prepare, run, stain/destain and quantitate one single SDS-gel.

An attempt to develop a generally applicable CE-method for controlling the consistency of glycoprotein production batches through monitoring the reproducibility of the glycoform pattern was made by Bonifichi [86]. He used a conventional Durawax fused-silica open tubular capillary GC column of 50 µm ID, internally coated with a 0.1 µm thick layer of DB-Wax stationary phase. The efficiencies obtained with this column were generally lower than 100'000 and the separation conditions with buffer composition, pH and field strength varied for each glycoprotein investigated. In general, low ionic-strength buffers with a pH far enough from the pI of the glycoprotein were found to be useful. This was demonstrated by the analysis of RNAse B, Ovalbumin and human and bovine α_1-acid glycoprotein. A method for the preparation of versatile hydrolytically stable cellulose-derivative coatings was recently introduced by Huang et al. [87]. These coatings which generate only low EOF at alkaline pH, were demonstrated to be hydrolytically stable in a pH range of 2–10 for several weeks.

As an example for basic glycoproteins, RNAse A and B were separated using a 25 mM Tris/ HCl buffer at acidic pH. Ovalbumin, as an example for acidic glycoproteins, was only parti- ally resolved in an alkaline 2-amino-2-methyl-1,3-propanediol/H$_3$PO$_4$ buffer.

5.4.1.12 CE-MS of glycoforms

Using on-line CE-MS coupling, glycoprotein glycoforms can be detected with high resolu- tion. The most promising ionization method for such an on-line approach was demonstrated to be electrospray ionization (ESI) [88]. Electrophoretic conditions optimized for glycoform resolution were found to be entirely compatible with the operation of electrospray mass spectrometry (ESMS) [89]. These separation conditions comprised e.g. acidic BGE's and Polybrene coated capillaries and allowed for the analysis of RNAse B, ovalbumin, horse- radish peroxidase and a lectin from *Erithrina corallodendrom*. The glycoform distribution obtained with CE-ESMS was virtually indistinguishable from those obtained with CE-UV as illustrated in Figure 5.19. High mannose structures containing glycoproteins other than RNAse B, such as the recombinant human bone morphogenetic protein-2, were also investi- gated by CE-ESMS [90]. The CE-ESMS approach was found to be a useful qualitative or even semiquantitative tool for comparing carbohydrate contents among different glycopro- teins, among isoforms of a given protein, or in batch-to-batch comparison of biopharmaceu- ticals. On-line coupling of CIEF and ESMS results in a two-dimensional separation system, where in the first dimension the samples are separated according to their pI and in the

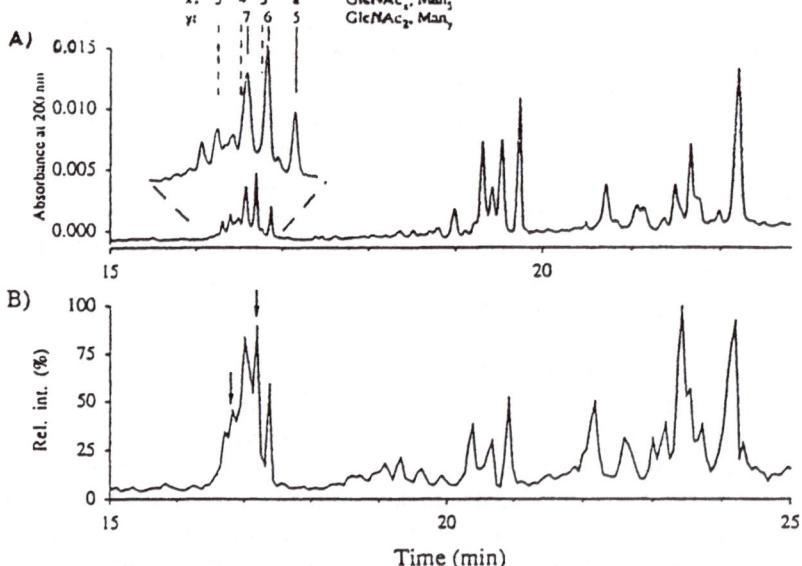

Figure 5.19 Analysis of peptides arising from CNBr cleavage of ovalbumin. (A) CE-UV analysis; (B) total ion electropherogram obtained for the CE-ESMS analysis (*m/z* 800–1800). Separation conditions: BGE, 2.0 M formic acid; capillary, 110 cm x 50 μm ID, coated with a solution of 5% Polybrene and 2 % ethylene glycol; injected amount, approx. 2 pmoles of ovalbumin digest (from [89], with permission).

second dimension according to their mass. On the basis of pI-differences, bovine serum apotransferrin glycoforms were separated into di-, tri- and tetrasialotransferrins [91]. Additional transferrin variants within each transferrin group, differing in their molecular weights, were then easily distinguished by ESMS. Another MS mode coupled to CE but in the off-line mode is matrix assisted laser desorption ionization time-of-flight mass spectrometry (MALDI-TOF-MS). After fraction collection of the separated ovalbumin and Desmodus salivary plasminogen activator glycoforms, the fractions were 1:1 diluted with the UV-absorbing matrix sinapinic acid and further characterized by MALDI-TOF-MS [92].

5.4.2 Glycopeptides

The separation of peptides was one of the first examples to demonstrate the high resolving power and selectivity of CE. However, only a few papers focus on glycopeptides, separated by CE based on differences in their carbohydrate moieties. Nashabeh et al. [93] demonstrated the CZE mapping of the tryptic peptide fragments of human AGP and the submapping of its glycosylated and nonglycosylated fragments. Prior to the CZE separation, the whole digest was fractionated into peptide and glycopeptide fragments by affinity chromatography on a concanavalin A (Con A) column, resulting in a Con A non-reactive, a Con A slightly reactive and a Con A reactive fraction. The CZE submapping of the Con A reactive peptides produced peaks that are missing from the submaps of all other fractions but appear in the whole map, thus allowing for the assessment of glycosylated fragments in the whole peptide map. High resolution and reproducible separations of rtPA glycopeptides carrying hybrid and complex type glycan chains were achieved in phosphate or putrescine containing tricine buffers [94]. The glycopeptides were obtained through tryptic digestion of the reduced and S-carboxymethylated glycoprotein and subsequent purification by HPLC resulting in fractions containing either the glycopeptide with the asn-448 site (glycopeptide 3) or the two glycopeptides with the asn-184 (glycopeptide 2) and the asn-117 site (glycopeptide 1). The combination of enzymatic digestion, HPLC fractionation and CE separation allowed for glycosylation site specific mapping of the rtPA glycopeptides (Figure 5.20). The combination of CE and HPLC was also used for monitoring the glycosylation of the two peptides dalargin and desmopressin [95]. Application of both techniques permitted the identification and determination of the reaction products as well as the determination of the reaction rate.

The concept of ion-pairing was demonstrated to be suitable in CE to evaluate the glycopeptide microheterogeneity of r-HuEPO expressed in CHO cells [96]. Using an acidic phosphate BGE with heptanesulfonic acid as the ion pairing agent, the tryptic fragments were separated into two regions, one containing the non-glycosylated and one containing the glycosylated peptides. Initial fractionation of specific asialoglycopeptides, produced through sialidase treatment of the tryptic glycopeptides by reversed phase HPLC (RP-HPLC), allowed for the CE mapping of the individual O- and N-glycosylation sites of r-HuEPO. As illustrated in Figure 5.21, the structural complexity of the asialoglycopeptides seem to increase from the simplest O-linked form (trace A) to the more complex asn-83 and asn-24 + asn-38 sites (trace B + C).

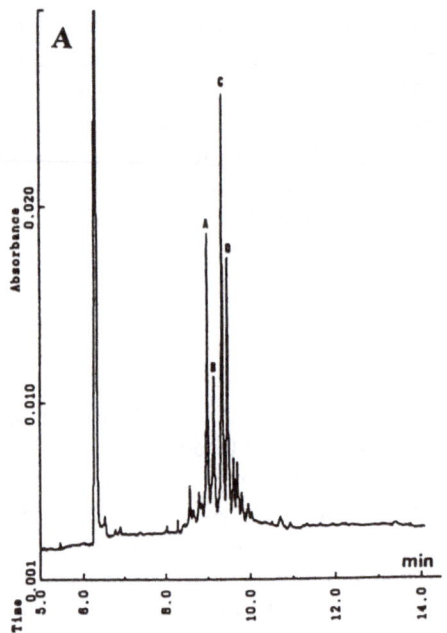
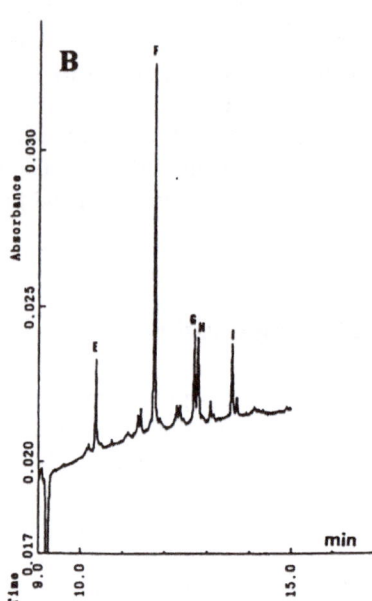

Figure 5.20 Analysis of rtPA (A) glycopeptide 2 and (B) glycopeptide 3 by CZE. Separation conditions: (a) BGE, 100 mM phosphate, pH 6.6; capillary, 57/50 cm x 75 µm ID; voltage,
10 kV; temperature, 40°C; UV detection, 200 nm; (b) BGE, 50 mM tricine, pH 8.5 with
2.5 mM putrescine; capillary, 107/100 cm x 75 µm ID; voltage, 30 kV; temperature,
30°C; UV detection, 200 nm (from [94], with permission).

Using a pH 7 phosphate BGE and UV detection, CE was applied to examine the dimerization constants of glycopeptide antibiotics, namely vancomycin, ristocetin and LY264826B
[97]. This study was performed since it is assumed that the dimerization is important in the
antibacterial activity of glycopeptide antibiotics. With CE, significant differences in the dimerization constants of the three antibiotics were revealed, following in general the same
trends as observed using nuclear magnetic resonance spectroscopy and sedimentation equilibrium.

Pulsed amperometric detection (PAD) which was already demonstrated to be suitable to
detect mono- and oligosaccharides after CE separation [12,16], can also be applied to glycopeptides. Four glycopeptide fractions, containing 0, 1, 2 or 4 sialic acid residues, were isolated from a tryptic digest of recombinant coagulation factor VII by RP-HPLC and analyzed
by CE using UV and PAD detection [98]. Although PAD detection usually requires the use
of a very high pH BGE (pH > 12), the four glycopeptides could be analyzed using a pH 9
borate buffer. CE-PAD was found to be specially suitable for the analysis of the PNGase F
treated glycopeptide fractions since it allows for the simultaneous detection of the peptide
backbone and the released oligosaccharides which is not possible with UV detection. CE-
PAD was also applied to the analysis of a tryptic digest of rtPA [99]. The use of different
detection potentials in sequential runs on the same sample gave structural information on the
peptides, such as glycosylation.

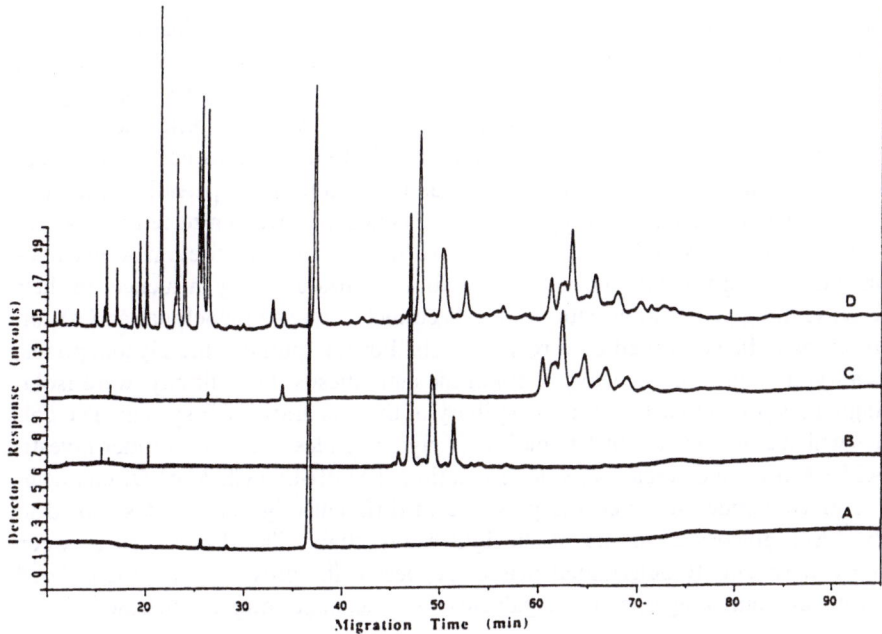

Figure 5.21 Evaluation of the asialo microheterogeneity associated with each glycosylation position of rHuEPO by HPCE mapping the reverse phase collected samples. (Trace A): O-linked site, (trace B): N83 site, (trace C) N24 + N38 site, (trace D) asialo rHuEPO total peptide control map. Separation conditions: BGE, 40 mM sodium phosphate, pH 2.5, containing 100 mM heptanesulfonic acid ion pairing agent; capillary, 75/50 cm x 50 μm ID; voltage, 16 kV; temperature, 30°C; UV detection, 200 nm. The sample was dissolved in 10% glacial acetic acid (from [96], with permission).

A combination of electrophoretic, chromatographic and mass spectrometric techniques was used to evaluate the glycosylation site heterogeneity of glycopeptides in proteolytic digests of bovine AGP [100]. CE analysis of the glycoprotein digests yielded electropherograms in which non-glycosylated and glycosylated peptides could be differentiated. Additionally, CE was applied to fast and highly efficient screening of HPLC fractionated AGP fragments which were obtained from proteinase digestion. By subsequent MS or LC/MS analysis the individual glycopeptides were identified and it was demonstrated that the CE separation relied upon the heterogeneity with respect to branching and sialylation within each glycosylation site. A very similar two-dimensional separation approach was recently introduced by Wu [101] for the analysis of glycosylated peptides from rtPA. Fractions containing glycopeptides were collected from RP-HPLC and analyzed by CZE, where separation of the carbohydrate structural variants occurred based on sialic acid content and branching on the same peptide. Non-glycosylated peptides collected in the same RP-HPLC fraction were well resolved from the glycosylated variants. The specific peaks occurring in the electropherograms were identified by MALDI.

CE/MS off-line coupling was applied by Birdwell et al. [102] for the evaluation of glycopeptides proteolytically released from bovine prothrombin fragment 1. To determine the sialic acid content of the structurally distinct carbohydrates attached at asn-77 and asn-101,

bovine prothrombin fragment 1 was first cleaved with α-chymotrypsin to release two glycopeptides, each containing one of the two glycosylation sites. In a second step the number of
sialic acids contained by each carbohydrate structure was established by ESMS and by CE
analysis of the isolated glycopeptides before and after treatment with sialidase and PNGase
F. CE separation with subsequent mass determination by MALDI-MS allowed for the glycosylation analysis of a human monoclonal anti-Rhesus (D) antibody [51]. Anti-Rhesus (D)
antibodies are routinely administered to pregnant Rhesus negative women to reduce the occurrence of haemolytic disease of the Rhesus positive newborn due to foetomaternal allo-
immunization. Since the glycosylation of immunoglobulins has a strong influence on their
biological activity, the glycosylation pattern of cell culture derived monoclonal anti-D immunoglobulin needs to be controlled on a regular basis. For that purpose the glycopeptides,
derived from a tryptic digest of the heavy chain of the anti-Rhesus (D) antibody, were isolated from a single injection into a 100 μm ID capillary of the preparative CE system. The collected samples which contained about 10 pmol/μl of glycopeptides were then further investigated by MALDI-MS for molecular mass determination. The off-line CE-MALDI characterization of the glycopeptides revealed the presence of different oligosaccharides linked to
the unique N^{297}-S-T glycosylation site of the IgG heavy chain. The differences between
calculated and experimentally determined masses suggested the presence of a fucosylated
biantennary structure containing one or two galactose units as major oligosaccharide.

5.4.3 Complex Oligosaccharides

Analysis of the isolated complex oligosaccharide structures associated with glycoproteins
require their release from the polypeptide backbone. Several cleaving methods for N- and
O-linked oligosaccharides have been established based on chemical release by hydrazinolysis or enzymatic release by the action of endoglycosidases. For a detailed discussion of the
cleaving and isolation procedure see references [103, 104]. Table 5.6 shows a selection of
N-linked oligosaccharide structures which are discussed in the following sections.

5.4.3.1 UV detection of derivatized complex oligosaccharides

Using ovalbumin as a model protein, Honda et al. [60] demonstrated capillary electrophoretic separation of glycoprotein derived oligosaccharides, released by hydrazinolysis and
labeled with 2-aminopyridine after re-N-acetylation. The ovalbumin glycans were separated
in two different electrolytes, a pH 2.5 phosphate BGE and an alkaline borate BGE (Figure
5.22). While the acidic phosphate BGE could not resolve solutes having the same degree of
polymerization, borate complexation succeeded in revealing the structural differences of the
labeled glycans confirmed by the larger number of peaks occurring in the electropherogram.
The same dual separation mode was applied to develop a two-dimensional mapping scheme
for complex oligosaccharides [105]. The sialic acid residues were removed from the pyridylamino derivatized glycans by neuraminidase digestion for simplicity of data interpretation.
After investigating the migration profiles of fifteen complex type, eleven high mannose type
and six hybrid type oligosaccharides, their mobilities relative to that of glucose were plotted

Table 5.6   Structures of N-linked complex oligosaccharides

High Mannose Type

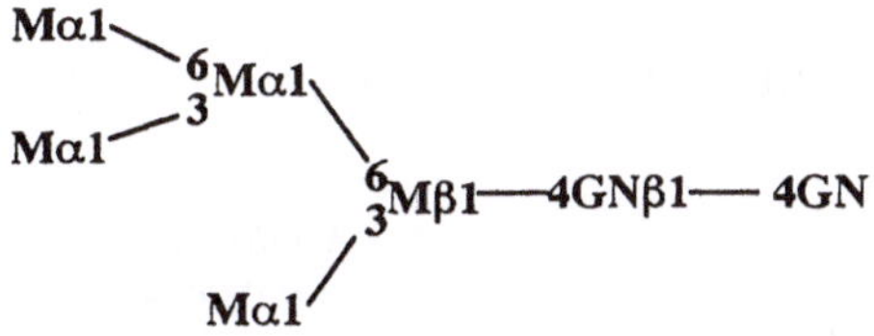

Mα1\
 ⁶Mα1\
Mα1/ ³   ⁶Mβ1——4GNβ1—— 4GN   **MAN5**   ($M_r = 1235$)

Mα1/ ³   *Oligomannose 5*

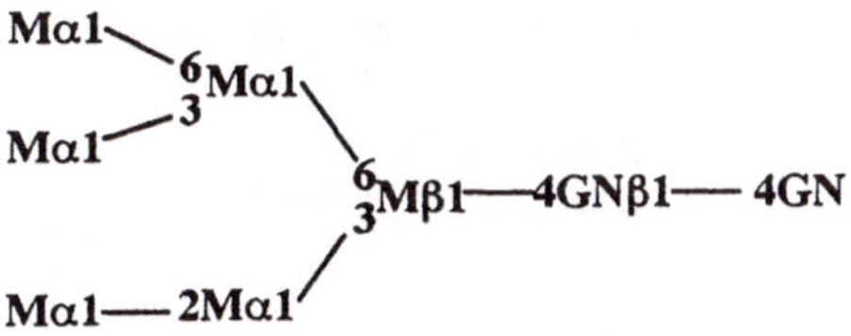

Mα1\
 ⁶Mα1\
Mα1/ ³   ⁶Mβ1——4GNβ1—— 4GN   **MAN6**   ($M_r = 1398$)

Mα1——2Mα1/ ³   *Oligomannose 6*

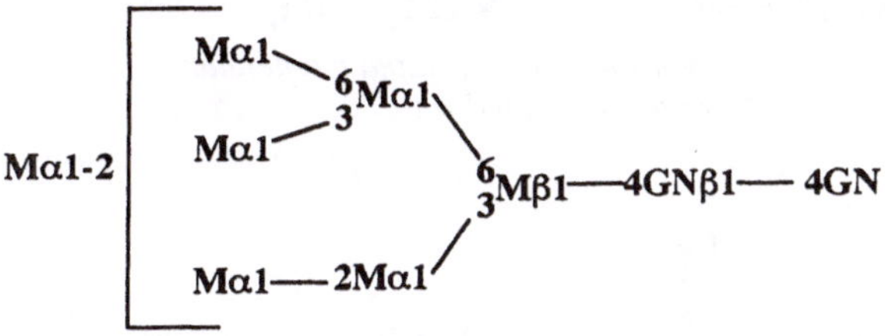

Mα1\
         ⁶Mα1\
Mα1/ ³
Mα1-2   ⁶Mβ1——4GNβ1—— 4GN   **MAN7**   ($M_r = 1560$)

Mα1——2Mα1/ ³   *Oligomannose 7*

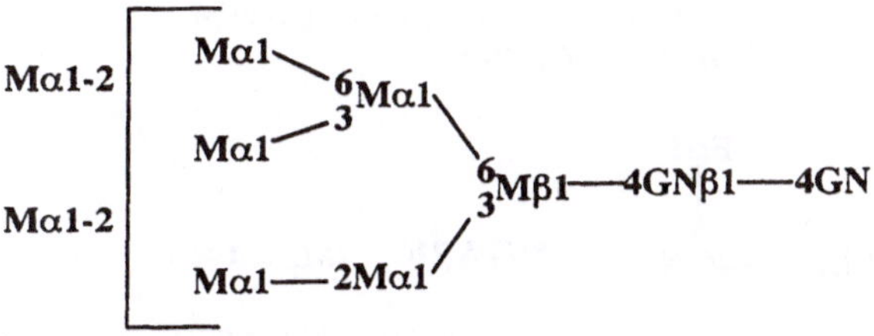

Mα1-2   Mα1\
         ⁶Mα1\
        Mα1/ ³
Mα1-2   ⁶Mβ1——4GNβ1——4GN   **MAN8**   ($M_r = 1398$)

Mα1——2Mα1/ ³   *Oligomannose 8*

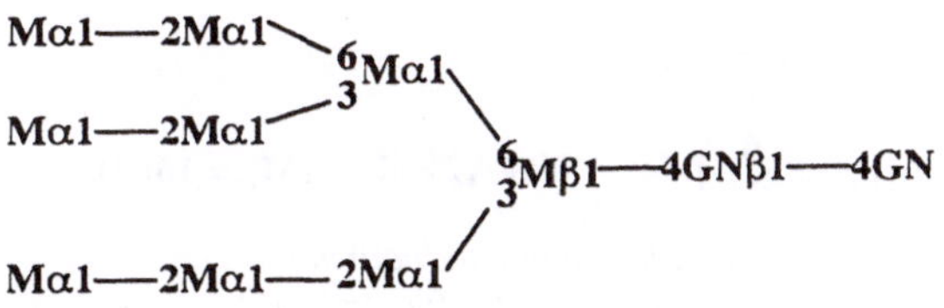

Mα1——2Mα1\
         ⁶Mα1\
Mα1——2Mα1/ ³
                     ⁶Mβ1——4GNβ1——4GN   **MAN9**   ($M_r = 1398$)

Mα1——2Mα1——2Mα1/ ³   *Oligomannose 9*

continue Table 5.6

Neutral Complex Type

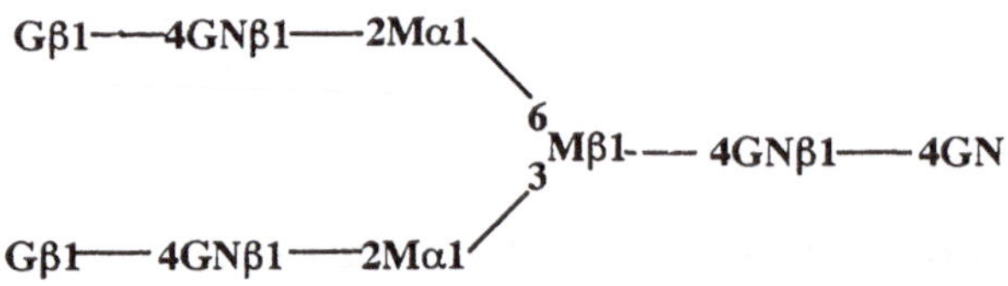

Gβ1——4GNβ1——2Mα1⟍
                        6
                        ⟍
                         Mβ1——4GNβ1——4GN   **NA2**   (M_r = 1642)
                        ⁄3
Gβ1——4GNβ1——2Mα1⁄

                                                  Asialo-,galactosylated
                                                  bi-antennary

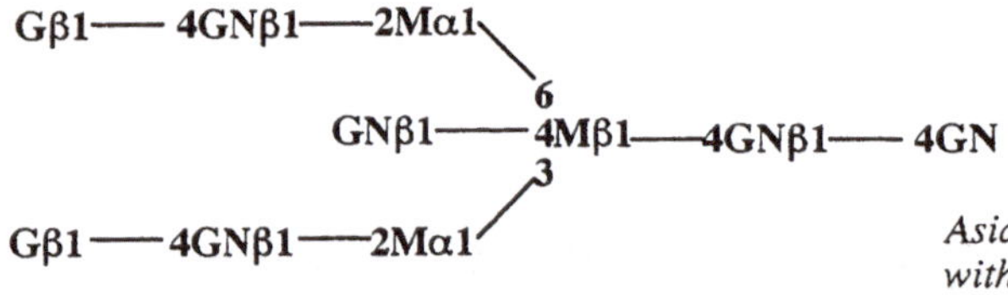

Gβ1——4GNβ1——2Mα1⟍
                        6
            GNβ1——4Mβ1——4GNβ1——4GN   **NA2B**   (M_r = 1845)
                        3
Gβ1——4GNβ1——2Mα1⁄

                                              Asialo-, galactosylated biantennary,
                                              with bisecting GlcNAc

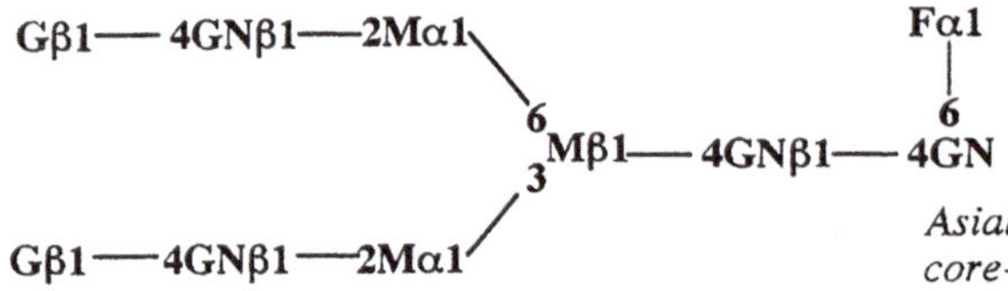

Gβ1——4GNβ1——2Mα1⟍   Fα1
                        6 |
                        ⟍   6
                         Mβ1——4GNβ1——4GN   **NA2F**   (M_r = 1788)
                        ⁄3
Gβ1——4GNβ1——2Mα1⁄   *Asialo-, galactosylated biantennary,*
                                               core-substituted with Fuc

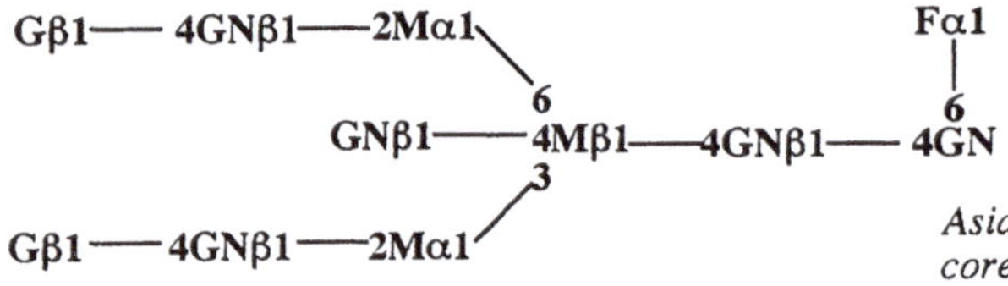

Gβ1——4GNβ1——2Mα1⟍   Fα1
                        6 |
            GNβ1——4Mβ1——4GNβ1——4GN   6
                        3   **NA2FB**   (M_r = 1991)
Gβ1——4GNβ1——2Mα1⁄
                                               Asialo-, galactosylated biantennary,
                                               core-substituted with Fuc and with
                                               bisecting GlcNAc

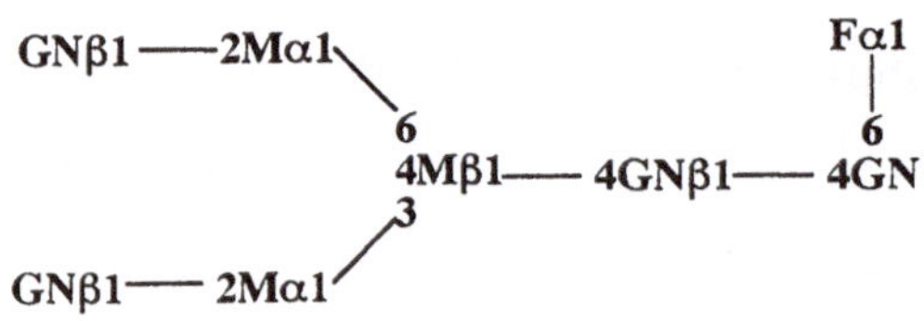

GNβ1——2Mα1⟍   Fα1
                   6 |
                4Mβ1——4GNβ1——4GN   6
                   3   **NGA2F**   (M_r = 1463)
GNβ1——2Mα1⁄
                                               Asialo-, agalacto-, biantennary,
                                               core-substituted with Fuc

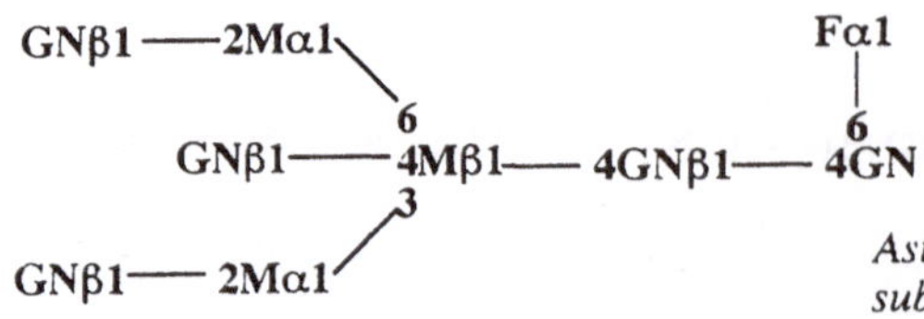

GNβ1——2Mα1⟍   Fα1
                   6 |
       GNβ1——4Mβ1——4GNβ1——4GN   6
                   3   **NGA2FB**   (M_r = 1667)
GNβ1——2Mα1⁄
                                               Asialo-, agalacto biantennary, core-
                                               substituted with Fuc and with
                                               bisecting GlcNAc

continue Table 5.6

Neutral Complex-Type

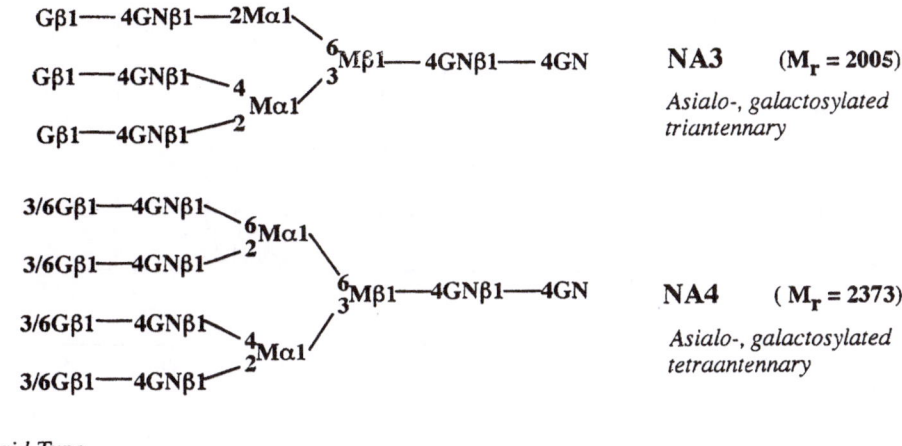

NA3 (M$_r$ = 2005)

Asialo-, galactosylated triantennary

NA4 (M$_r$ = 2373)

Asialo-, galactosylated tetraantennary

Hybrid-Type

HYBR (M$_r$ = 1642)

Hybrid-type with bisecting GlcNAc

Sialylated Complex- type

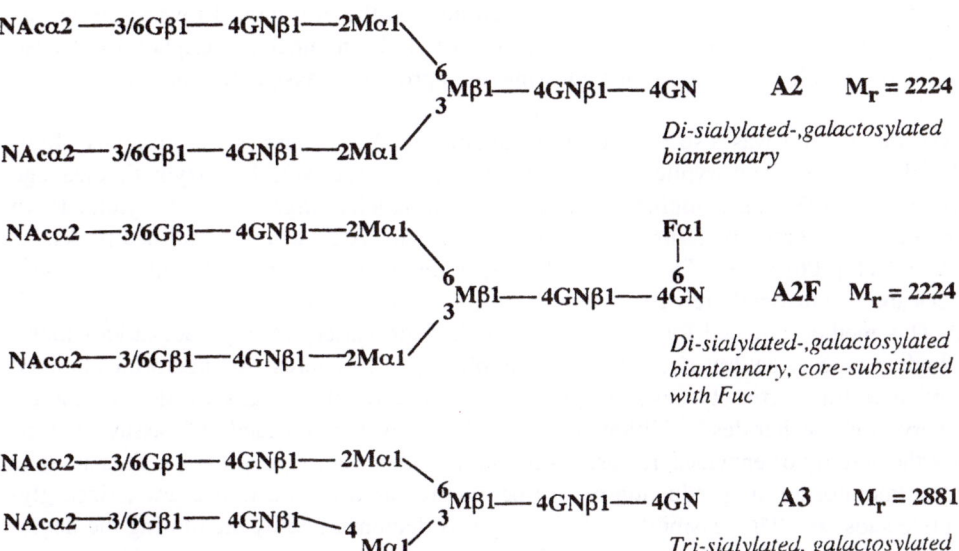

A2 M$_r$ = 2224

Di-sialylated-,galactosylated biantennary

A2F M$_r$ = 2224

Di-sialylated-,galactosylated biantennary, core-substituted with Fuc

A3 M$_r$ = 2881

Tri-sialylated, galactosylated triantennary

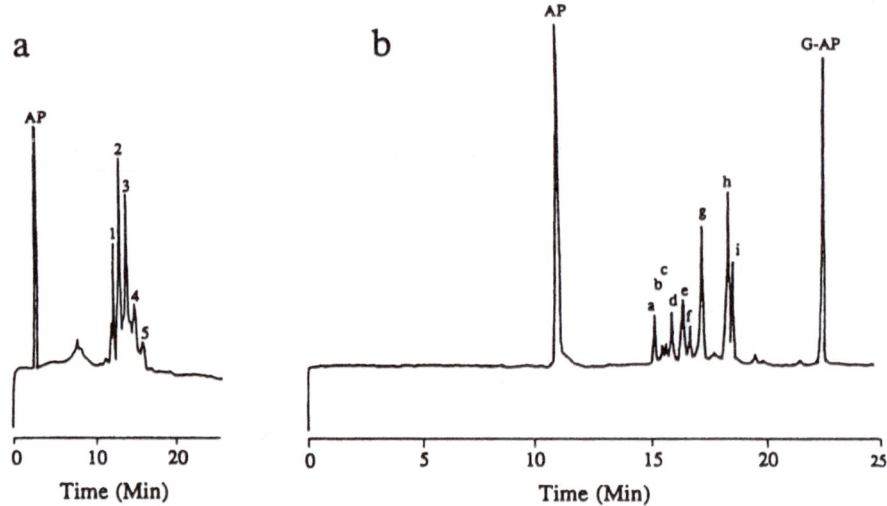

Figure 5.22 Analysis of reductively pyridylaminated oligosaccharides derived from ovalbumin (a) by direct CZE or (b) CZE as borate complexes. Separation conditions: (a) BGE, 0.1 M phosphate, pH 2.5; capillary, 20 cm x 25 µm ID; voltage, 8 kV, UV detection, 240 nm; (b) BGE, 200 mM borate, pH 10.5; capillary, 95 cm x 50 µm ID; voltage, 20 kV; fluorescence detection, excitation: 316 nm, emission: 395 nm. Peak assignment: 1 = heptasaccharide, 2 = octasaccharide, 3 = nonasaccharide, 4 = decasaccharide, 5 = undecasaccharide. g = HYBR, h = MAN6+7, i = MAN5 (for structures see Table 5.6), (from [60], with permission).

in a two-dimensional map. The oligosaccharides investigated were positioned in three distinct domains, one for each glycan type. In each domain the plot moved toward the left as the degree of polymerization increased and upward as the number of peripheral mannose and galactose residues increased thus allowing an approximate assignment of unknown glycans.

Pyridylamination was also proven to be suitable for the CE analysis of complex oligosaccharides derived from tryptic digests of human and bovine AGP by enzymatic cleavage with PNGase F [93]. CE mapping of the released and labeled oligosaccharides yielded two different electrophoretic patterns for human and for bovine AGP with well defined peaks.

Yuen et al. [106] reported a CE based fingerprinting method in which N-linked oligosaccharides are enzymatically released from the glycoprotein and subsequently derivatized with PMP. This method allowed for the separation of a wide variety of oligosaccharides including high mannose, neutral and sialylated complex type structures. A fingerprint obtained from released transferrin glycans showed baseline resolution of several sialylated tri- and biantennary oligosaccharides. Additionally, the authors developed a facile CE-assay for estimating the activity of enzymes, releasing oligosaccharides from glycoproteins. This assay is based on monitoring the production of free peptide released from a well characterized glycopeptide substrate. The feasibility of this assay was demonstrated by monitoring the deglycosylation of transferrin using various PNGase F concentrations.

5.4.3.2 UV detection of underivatized complex oligosaccharides

Sialylated oligosaccharides released from AGP by conventional hydrazinolysis can be ana-lyzed by CE with UV detection at 190 nm. This method does not require a derivatization or borate complexation, as demonstrated by Hermentin et al. [107]. The carbonyl function of the N-acetyl and carboxyl groups present in the sialylated oligosaccharides enabled their direct UV detection in the femtomole range. The same AGP oligosaccharides were analyzed by means of HPAEC-PAD, but with a sensitivity 4000 times less than in CE in terms of injected amount. In the same study the authors compared the suitability of chemical and enzymatic (PNGase F) cleaving procedures by means of the resulting N-glycan profiles. They found an optimized hydrazinolysis procedure to be the best with practically no loss of NeuAc and the closest resemblance to the PNGase F-derived N-glycan pool. In a subsequent paper, Hermentin et al. [108] further optimized this CE method in order to establish a carbo-hydrate mapping database. With mesityloxide and sialic acid as internal standards and by applying a triple correction method, accurate and highly reproducible migration times with a coefficient of variation of less than 0.20% were obtained. Approximately 80 different sial-ylated N-glycans of known structure have been entered into a Lotus 1-2-3 mapping data-base, which has been verified using N-glycans released from urinary rHuEPO, from bovine serum fetuin and from AGP. The efficiency of the database and the triple-correction method was further confirmed by interlaboratory CE measurements.

An alternative approach for the separation of glycoprotein derived glycans was shown to be MECC using SDS as the micellar phase. Taverna et al. [109] investigated the influence of various experimental parameters such as the concentration of surfactant and electrolyte and the addition of divalent cations on the separation of neutral and sialylated N-linked glycans released from rtPA by N-glycanase digestion. While neutral oligosaccharides were separated mainly on the basis of their solubilization in the micelles, for sialylated oligosaccharides the electrophoretic mobility, determined by the degree of sialylation, remained the predominant factor. This is due to the electrostatic repulsion between the micelles and the negatively charged sialic acid residues. The selectivity of the MECC system was further enhanced through addition of Mg^{2+} to the micellar phase. The improved selectivity and resolution under these conditions was a result of both an increase in the migration time window and the differential complexation of the carbohydrates with the divalent metal cation. It was assu-med that the complex stability mainly depends on the conformation of free hydroxyl groups in the carbohydrate chain and on the absence of terminal sialic acid residues. This improved MECC separation system was applied for the complete N-glycosylation mapping of rtPA [110]. rtPA oligosaccharides were first fractionated by anion exchange chromatography and then separated by MECC. As expected, the elution order of the oligosaccharides followed mainly the degree of sialylation starting with the neutral up to the tetrasialylated structures. However, within each group some separation was still attainable, indicating that the separa-tion relied not only on charge differences but also on structural differences between oligo-saccharides bearing the same number of sialic acid residues. Analysis of the N-glycans re-leased from the distinct glycopeptides obtained from tryptic digest of rtPA allowed for the site specific glycosylation mapping (Figure 5.23). The oligosaccharide mapping revealed that asn-448 and asn-184 carry the same population of complex type oligosaccharides but in varying relative amounts. HPAEC analysis of the desialylated oligosaccharides proved that

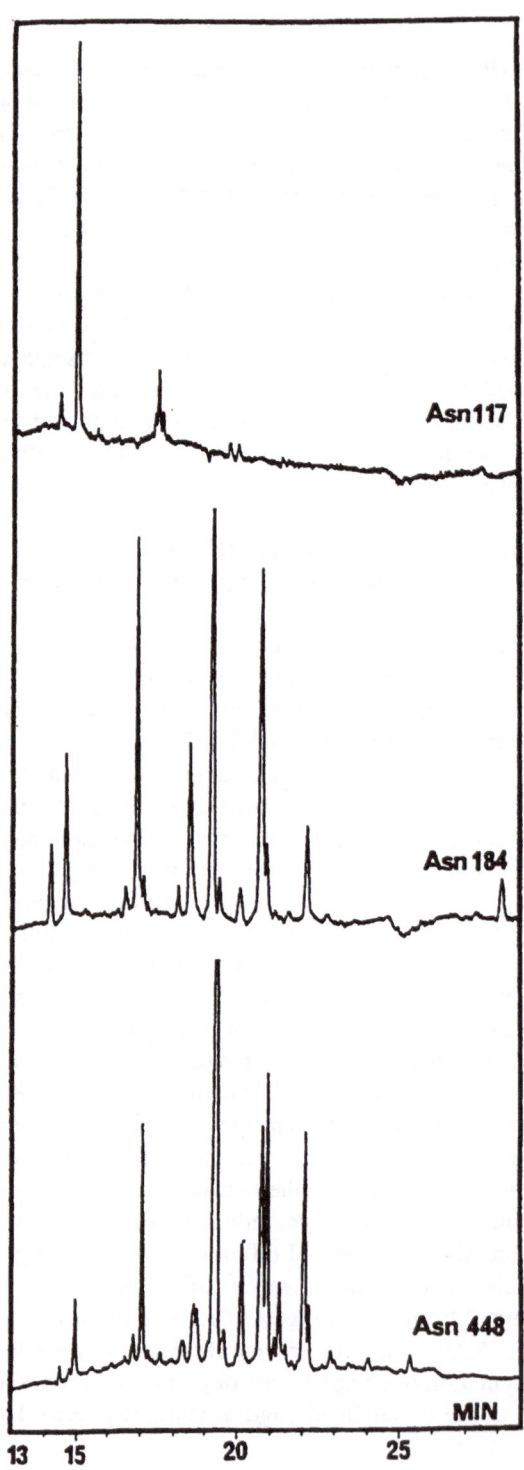

Figure 5.23 Oligosaccharide mapping of the three glycosylated sites of rt-PA: (a) Asn-117, (b) Asn-184, (c) Asn-448. Separation conditions: BGE, 50 mM phosphate – 10 mM MgCl$_2$, pH 7; capillary, 97/90 cm x 75 µm ID; voltage, 22 kV; temperature, 30°C; injection, 2 sec, 35 mbar; UV detection, 200 nm (from [110], with permission).

the Asn-448 site not only contains a greater proportion of heavily sialylated structures but also has a higher degree of microheterogeneity.

Evaluation of CE as a possibly complementary procedure to HPAEC-PAD in the assessment of the purity of oligosaccharide mixtures was the aim of another study [111]. Underivatized oligosaccharides were separated in a phytic-acid borate buffer system. The presence of the phytic acid as an ion pairing agent was assumed to reduce interaction with the capillary wall and to possibly complex with the solutes thus enhancing the selectivity. CE buffer conditions such as the pH, borate concentration and ion pairing agent concentration were varied to optimize the separation of complex oligosaccharides derived from IgG antibodies and CTLA4Ig fusion protein. Whilst the separation between neutral species was enhanced by an increased borate concentration no positive influence on the resolution of sialylated species was observed. The IgG and CTLA4 separations showed certain correlations with the corresponding HPAEC-PAD chromatograms. However, no attempts were made to assign distinct oligosaccharide structures to the peaks detected.

Treatment of glycoproteins with alkali in the presence of borohydride liberates O-glycosydically linked oligosaccharides as their corresponding alditols. Since they lack the aldehyde group as the key functional group for any chemical reaction these alditols are not accessible for derivatization. Therefore Honda and coworkers [112] developed an analysis scheme for oligosaccharide alditols using UV detection at 185 nm. In a SDS containing borate buffer several O-linked monosialooligosaccharides were separated and detected at the 10^{-4} M level. This was possible due to the enhanced UV absorption of the N-acetyl groups present in sialic acid and hexosamine residues. Addition of SDS to the separation buffer was required in order to resolve glycans differing only in containing a N-glycolylneuraminic acid (NeuGc) residue instead of a N-acetylneuraminic acid (NeuAc) residue (for structures see Figure 3.6). Separation of an equimolar mixture of sialooligosaccharide standards showed that NeuGc containing glycans (peak 2+4 in Figure 5.24) exhibit a slightly longer migration time compared to the NeuAc containing glycans (peak 1+3 in Figure 5.24). The established system was applied to the microanalysis of sialooligosaccharides in bovine submaxillary mucin and swallow nest material.

5.4.3.3 Laser induced fluorescence detection (LIF) of complex oligosaccharides

2-Aminoacridone (AMAC) and 8-aminonaphthalene-1,3,6-trisulfonic acid (ANTS) were applied as derivatization agents for developing electrophoretic separation schemes for neutral and sialylated complex oligosaccharides based on slab gel electrophoresis [113, 114]. Recently, both agents were also successfully introduced in highly efficient and sensitive CE based analysis schemes [115,116].

8-Aminonaphthalene-1,3,6-trisulfonic acid (ANTS)

Low nanomole and picomole amounts of high mannose and complex type oligosaccharides were labeled with ANTS in a total reaction volume of as little as 2 µl and subsequently detected using He-Cd laser induced fluorescence detection [115]. In case of homologous high mannose type oligosaccharides derived from RNAse B, separation in a pH 2.5 phosphate BGE was found to be based on differences in the charge-to-mass ratio of the solutes. Since

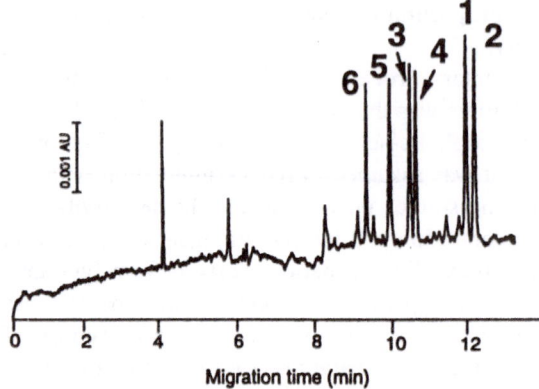

Figure 5.24 Separation of an equimolar mixture of sialooligosaccharide alditol standards. Separation conditions: BGE, 200 mM borate, pH 9.6, containing 0.1 M SDS; capillary, 50 cm x 50 µm ID; voltage, 17 kV, UV detection, 185 nm (from [112], with permission).

the oligomannoses are neutral carrying only the negative charges from the ANTS label, they migrated in the order of increasing molecular weight from MAN5 to MAN9, as illustrated in Figure 5.25 (for structures see Table 5.6). This regular migration behavior was confirmed by a plot of electrophoretic mobilities versus charge-to-mass ratio of the oligomannoses which yielded a linear relationship (see Figure 4.15). Investigation of a series of complex type oligosaccharides differing in size and charge revealed that again the principal separation mechanism relied on the differences in the charge-to-mass ratios. However, the strict linear electrophoretic mobility versus charge-to-mass ratio relationship could not be applied (see Chapter 4.2.2.1).

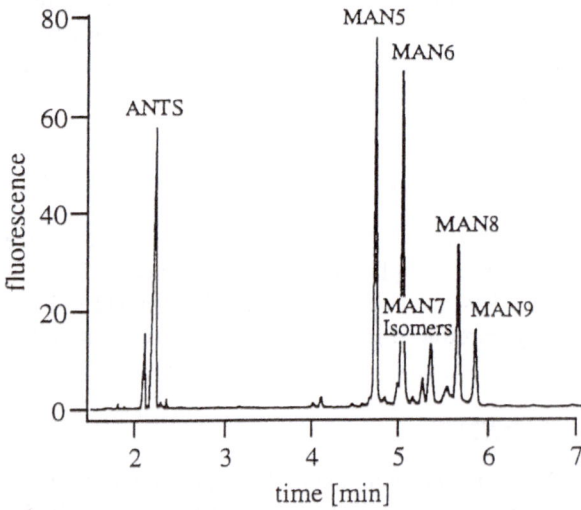

Figure 5.25 Separation of ANTS labeled oligomannose-type oligosaccharides MAN5 – MAN9. Separation conditions: BGE, 50 mM phosphate, pH 2.5; capillary, 27/20 cm x 50 µm ID; voltage, –10 kV; temperature, 25°C; LIF detection, He-Cd laser, excitation: 325 nm, emission: 520 nm. Injected sample concentration, 250 µg/ml. For structures of MAN5 – MAN9 see Table 5.6 (from [115], with permission).

That ANTS derivatization allows for the fast generation of oligosaccharide patterns and is suitable to reveal distinct differences between the oligosaccharide composition of different glycoproteins is demonstrated in Figure 5.26 with the separation of the glycan pools derived from human immunoglobulin G (IgG) and bovine fetuin [117].

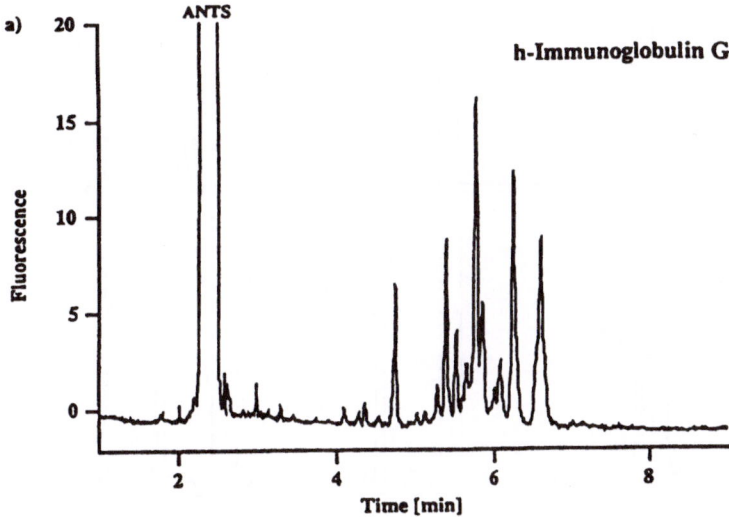

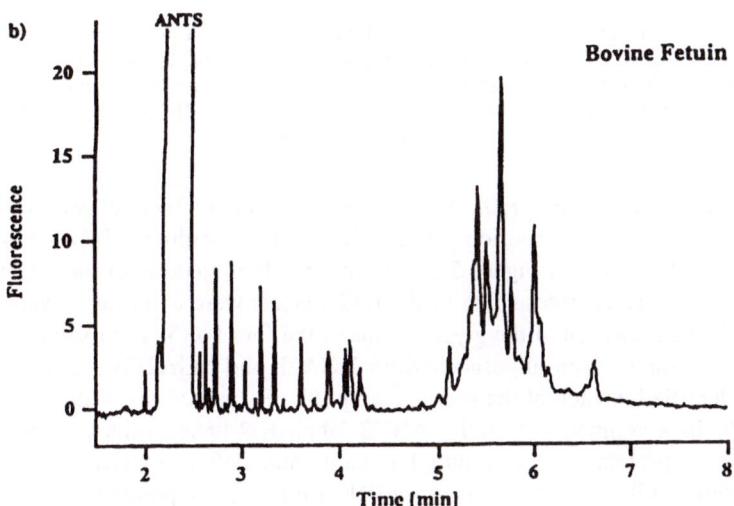

Figure 5.26 Separation of ANTS-labeled oligosaccharide libraries derived from a) human immunoglobulin G and b) bovine fetuin. CE conditions: BGE, 50 mM phosphate, pH 2.5; capillary, 27/20 cm x 50 μm ID; voltage, –10 kV; temperature, 25°C; injection, 3 sec at the cathodic end; LIF detection, He-Cd laser, excitation: 325 nm, emission: 520 nm. (from [117], with permission).

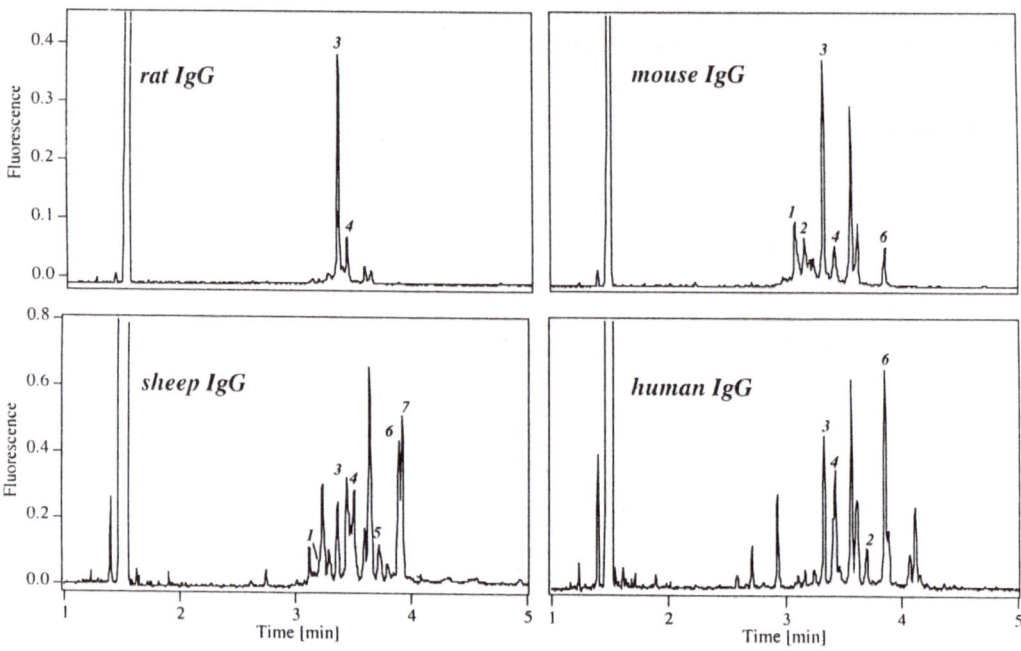

Figure 5.27 Comparison of the oligosaccharide patterns of IgG derived from four different sources. Separation conditions: BGE, 50 mM phosphate, pH 2.5; capillary, 27/20.5 cm x 50 µm ID; voltage, −10 kV; temperature, 25°C; LIF detection, He-Cd laser, excitation: 325 nm, emission: 520 nm. Peak assignment: 1 = A2, 2 = A2F, 3 = NGA2F, 4 = NGA2FB, 5 = NA2, 6 = NA2F, 7 = NA2FB (for structures see Table 5.6).

The high resolving power of the low EOF acidic phosphate system even allowed to reveal distinct differences in the oligosaccharide patterns of immunoglobulin G (IgG) derived from different sources, as illustrated in Figure 5.27. By means of coinjection of the isolated ANTS labeled complex oligosaccharides some of the IgG glycans were identified revealing certain similarities in the composition of the glycan pools of the four IgG's. For example the two neutral biantennary complex type oligosaccharides NGA2F and NGA2FB (peak 3 + 4 in Figure 5.27) were identified in each of the four glycan pools. For structures of the IgG N-glycans see Table 5.6. In a comparative study, ANTS labeled N-linked oligosaccharides released from several glycoproteins such as fetuin, human immunodeficiency virus envelope recombinant glycoprotein, AGP, and RNAse B using PNGase F, were separated by CE and polyacrylamide slab gel electrophoresis [118]. With both methods high resolution separations were easily attained. While in CE due to the high theoretical plate number excellent resolution was obtained without the use of a sieving matrix, in slab gel PAGE the separation relied on both the different charge-to-mass ratios of the solutes and the sieving effect. From these results it was concluded that the two methods are complementary and each offers unique advantages.

2-Aminoacridone (AMAC)

For the analysis of the neutral AMAC-carbohydrate conjugates, a MECC system was developed based on a high concentrated borate buffer containing taurodeoxycholate and laser induced fluorescence detection [116]. Due to its hydrophobicity excess AMAC was completely entrapped in the slow migrating micelles thus not interfering with the analysis. The migration order of the AMAC labeled high mannose type oligosaccharides derived from RNAse B in the micellar system was reversed compared to that observed for the same solutes labeled with ANTS and separated under low EOF conditions and reversed polarity (Figure 5.25). While the ANTS derivatives were separated within 6 min., the separation of the AMAC derivatives required almost 30 min., obtaining similar resolution. The authors attempted to establish a quantitative relationship between the mobility of the AMAC-conjugates and the hydrodynamic volume of the oligosaccharides which is related to the three dimensional structure of the monosaccharide residues. The plot of the mobilities versus the hydrodynamic volume, determined by adding all the contributions from each sugar subunit expressed relative to that for glucose, showed a linear relationship for both for the linear and the branched oligosaccharides investigated. This concept was applied to tentatively assign unknown peaks in an ovalbumin derived oligosaccharide pattern to distinct structures.

The present MECC method was also applied for fingerprinting AMAC labeled complex oligosaccharides released from bovine fetuin and human IgG monoclonal antibodies before and after treatment with neuraminidase [119]. An optimized automated hydrazinolysis procedure allowed to selectively release and analyze N- and O-linked glycans from bovine fetuin as illustrated in Figure 5.28. The fact that the N-linked glycans showed shorter migration times than the corresponding O-linked material indicated that N-linked glycans are usually larger and more complex in structure. To correlate data obtained at different times it was required to run a dextran ladder as a calibration standard to check consistency of migration times and resolution. MECC of AMAC labeled core fucosylated glycans before and after fucosidase digestion allowed for the general differentiation of fucosylated and non-fucosylated structures [120]. The fucosylated glycans were found to migrate consistently slower than the corresponding oligosaccharides lacking the core fucose. This was attributed to differences in the complex formation with borate and/or the micellar medium rather than to differences in their hydrodynamic volumes. This methodology was applied to obtain preliminary information on the carbohydrate composition of a monoclonal IgG antibody and the soluble complement receptor type 1. The hydrophobic character of the AMAC derivatives allowed a combined approach of RP-HPLC, MECC and off-line MS to characterize oligosaccharides [121]. AMAC labeled dextran and N-linked mannose rich glycan samples were first separated and concentrated by RP-HPLC. After collecting the heterogeneous HPLC fractions they were further separated by MECC and the results confirmed by MS. In both separation techniques the elution order of the labeled oligosaccharides was found to be related to the size of the molecules, indicating that hydrophobicity plays a major role in the separation mechanism.

8-Aminopyrene-3,6,8-trisulfonic acid (APTS)

Another separation mode, capillary gel electrophoresis (CGE) was applied by Guttman et al. [122, 123] to separate N-linked oligosaccharides derived from bovine pancreatic RNAse B and bovine fetuin. After enzymatic release, the glycans were labeled by reductive amination

with the fluorescent APTS and resolved in a neutrally coated capillary using a 25 mM acetate buffer, pH 4.75 which contained 0.4% polyethyleneoxide (PEO). A buffer pH of 4.75
enhanced the resolution specifically in case of the sialylated fetuin glycans since at that pH
all the sialic acids are ionized. By means of coinjection of individual complex oligosaccharide standards the high mannose structures of RNAse B and two of the sialylated structures
in the bovine fetuin library were identified. This CGE separation system even allowed for
the baseline resolution of the three positional isomers of MAN7 and MAN8 (Figure 5.29).
Ferguson plots of the logarithms of the electrophoretic mobilities versus the PEO concentration in the running buffer showed no apparent sieving effect for the APTS labeled high
mannose and sialylated complex type oligosaccharides. It was assumed that separation of the
labeled glycans in CGE relies on differences in the charge-to-mass ratios and the hydrodynamic volumes of the branched solutes rather than on specific sieving effects, as it was

a)   **N+O Linked Fetuin Oligosaccharide Library**

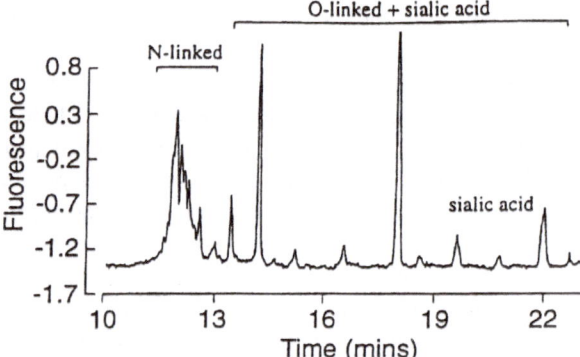

b)   **O Linked Fetuin Oligosaccharide Library**

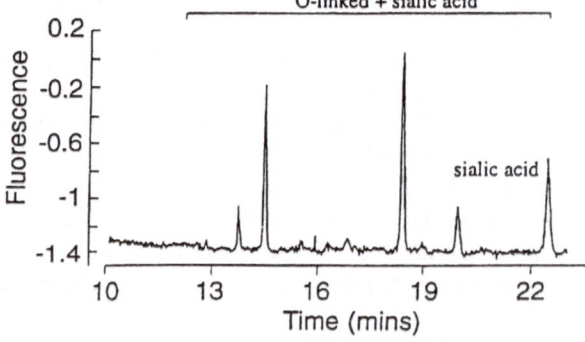

Figure 5.28 Electropherograms showing 2-aminoacridone derivatives of (a) the N- and O-linked
glycan library and (b) the O-linked glycan library from bovine fetuin. Separation conditions: BGE, 500 mM borate with 80 mM taurodeoxycholate, pH 8.9; capillary, 57 cm x
50 μm ID; voltage, 25 kV; temperature, 25°C; LIF detection, He-Cd laser, excitation:
442 nm, emission: 525 nm (from [119], with permission).

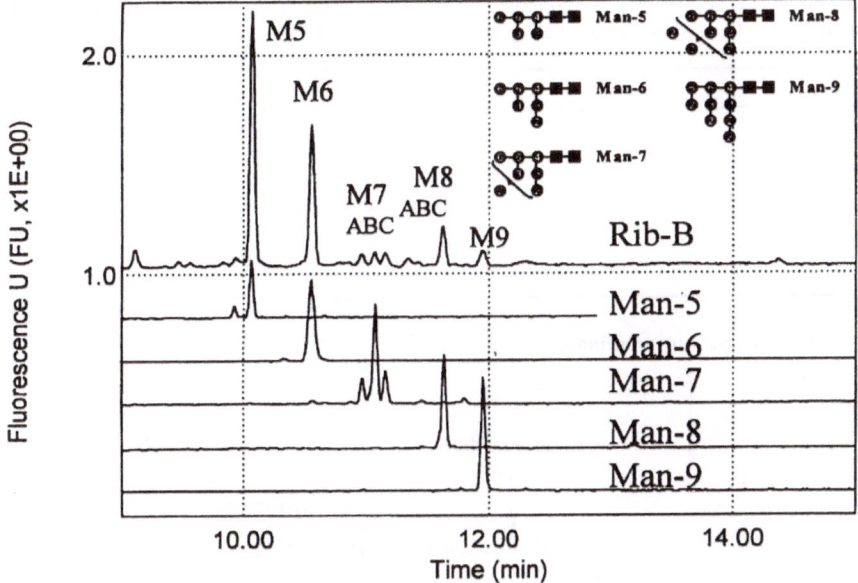

Figure 5.29 CGE separation of the APTS-labeled high mannose type oligosaccharides released from bovine ribonuclease B (upper trace) and the individual standard structures (lower traces). Inset: Structural representation of the high mannose type N-linked oligosaccharides: ■ = GlcNAcβ1,4; ◑ = Manβ1,4; ◎ = Manα1,6; ◕ = Manα1,3; ◔ = Manα1,2. Separation conditions: BGE, 25 mM acetate, pH 4.75, containing 0.4% PEO; capillary, 57/50 cm x 50 μm ID, neutrally coated; voltage, 500 V/cm (19 μA); temperature, 20 °C; LIF detection, argon-ion laser, excitation: 488 nm, emission: 520 nm (from [122], with permission).

demonstrated by Klockow et al. [115] for CZE separation systems. In the case of the high mannose glycans, coinjection with a maltooligosaccharide ladder standard provided the possibility of accurate assessment of the corresponding glucose unit (GU) values of the individual structures [124]. It additionally allowed the assessment of the contribution of each Man-α-1,2-residue to the migration characteristics of each solute. GU values could be calculated very precisely either for the different size structures or the same size positional isomers (for GU calculations see text below). The same approach was used to assign distinct structures in the bovine fetuin library [123]. According to their calculated GU values the two unknown major peaks were assigned to tetrasialylated triantennary structures.

Applying an array of exoglycosidases on oligosaccharides released from the glycoprotein allowed for oligosaccharide sequencing by consecutive enzymatic digestion followed by CE separation [125]. Multistructure sequencing of a complex glycan pool was achieved using enzyme mixtures of neuraminidase, β-galactosidase, β-N-hexosaminidase and α-mannosidase in different compositions. The composition of these enzyme mixtures and the cleaving sites in the glycan are indicated in Table 5.7. The first step involved APTS labeling of the reductive ends of all carbohydrates and the second step the enzymatic digestion of the labeled solutes with a specially designed exoglycosidase mix. The resulting fragments were

Table 5.7   Carbohydrate sequencing exoglycosidase enzyme assay. The lower panel depicts the cleavage spots of the individual enzymes in the matrix [125].

Enzymes / vials	1	2	3	4	5
Neuraminidase	x	x	x	x	x
β-Galactosidase	–	x	x	x	x
β-N-Acetylhexosaminidase	–	–	x	x	x
α-Mannosidase	–	–	–	x	x
α-Fucosidase	–	–	–	–	x

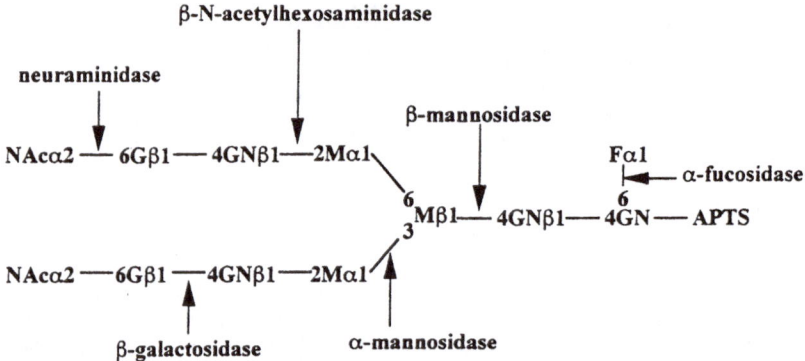

analyzed after each individual digestion step by CE-LIF. By comparison of the migration times of the exoglycosidase digest fragments to those of a maltooligosaccharide ladder the migration time shifts due to the enzymatic cleavage were determined. The glucose unit (GU) values of the various carbohydrate structures and their digestion fragments were calculated by

$$GU_x = G_n + \frac{MT_x - MT_n}{MT_{n+1} - MT_n}$$

where GU is the glucose unit value of the peak of interest, G_n is the degree of polymerization of the maltooligosaccharide unit immediately preceding the peak of interest, MT_x is the migration time of the peak of interest, and MT_n and MT_{n+1} are the migration times of the maltooligosaccharides immediately preceding and following the peak of interest, respectively. Comparison of the calculated values with tabulated GU values (Table 5.8) of the various monosaccharide units and core structures in glycoprotein derived oligosaccharides [125], allowed for the identification of distinct peaks generated by the exoglycosidase digestion. Additionally, comparison of the migration time shifts between undigested and digested structures helped to determine the unknown oligosaccharide sequences. In a further study this exoglycosidase mediated sequencing approach was applied to multistructure sequencing of N-linked fetuin glycans [126].

    In addition to the identification of unknown oligosaccharide structures through a step-by-step enzymatic digest, APTS labeled oligosaccharides were also characterized by off-line

Table 5.8 Glucose unit values (GU) of various monosaccharide units and core structures [125]

Core structures / monosaccharides	GU
APTS-GlcNAc$_2$Man (trisaccharide core)	3.21
APTS- GlcNAc$_2$Man$_3$ (N-linked core)	5.00
Man	0.9
GlcNAc	1.05
Gal	1.2
Fuc	0.7
Neu5Ac	−2.7

coupling of CE and MALDI-TOF MS [127]. First the APTS-conjugates were separated by CE and isolated using an automated high-resolution fraction collector. In a second step the isolated oligosaccharides were analyzed and detected by MALDI-TOF MS in the negative ionization mode, using a matrix consisting of (1:1) 6-hydroxypinolinic acid and 3-hydroxypinolinic acid. Sample clean-up was performed on-probe by means of a cation-exchange resin. Detection limits found for APTS-derivatized maltoheptaose as a standard oligosaccharide were around 30 femtomoles. Various standard oligosaccharides as well as the oligosaccharide pool derived from ribonuclease B were separated and collected by CE, followed by molecular mass determination with MALDI-TOF MS. It was found that for multisialylated species, such as tri-, tetra-, pentasialo and higher species, the MALDI-TOF procedure is not as sensitive as for non-, mono-, or bisialo APTS derived carbohydrates.

5.5 Glycosaminoglycans

Except hyaluronic acid all glycosaminoglycans (GAG's) are sulfated and therefore negatively charged. Due to their large size and structural complexity GAG's are usually depolymerized by the action of site specific enzymes before they are subjected to analysis and structural determination. Chondroitinases and heparinases function as eliminases and cleave the glycosidic linkage between the reducing end of an amino sugar and the hydroxyl group of the neighboring uronic acid. The final products are disaccharide units comprising the Δ^4-uronic acid attached to GlcNAc or GalNAc with varying degree of sulfation (Figure 5.30a and b). Whilst chondroitinase-ABC reacts with all members of the chondroitin/dermatan sulfate group and also with hyaluronan, chondroitinase-AC eliminates only aminosugars attached to GlcA and chondroitinase-B eliminates aminosugars attached to IdoA [128]. The three heparin lyases I, II and III also show different specificities, with the heparin lyase II and III being relatively nonspecific [129]. Hyaluronan might be digested very specific with testicular hyaluronidase which cleaves the GlcNAc-GlcA-linkage, forming oligomers with the GlcNAc as the reducing terminus as shown in Figure 5.30c [130].

The first applications of CE to the analysis of GAG's were presented by Al Hakim and Linhardt [131] and Carney and Osborne [132] using a basic borate buffer. Exhaustive treata)

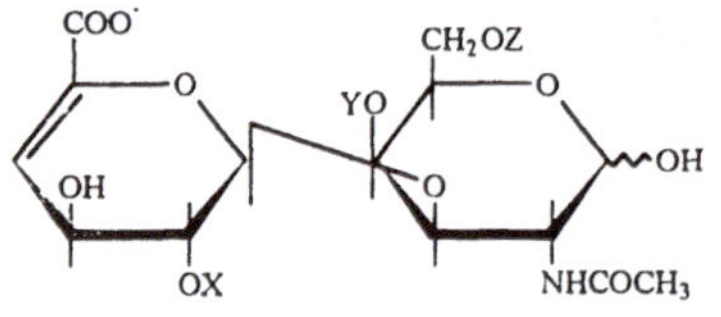

2, X = Y = Z = H; 3, X = Y = H, Z = SO$_3^-$ (Δ-6S);
4, X = Z = H, Y = SO$_3^-$ (Δ-4S); 5. Y = Z = H, X = SO$_3^-$ (Δ-2S);
6, Y = Z = SO$_3^-$, X = H; 7, X = Z = SO$_3^-$, Y = H;
8. X = Y = SO$_3^-$, Z = H; 9, X = Y = Z = SO$_3^-$.

Chondroitin Sulfate/Dermatan Sulfate Disaccharides

b)

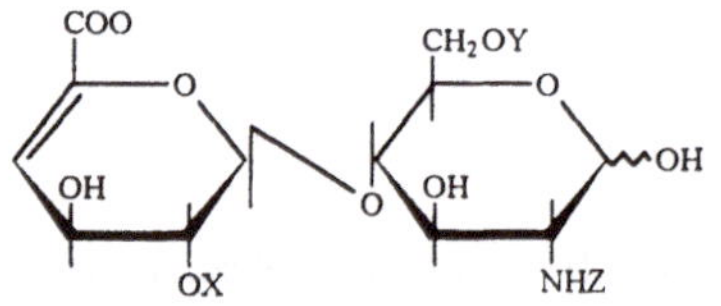

1h, X = Y = H, Z = Ac; 2h, X = H, Y = SO$_3^-$, Z = Ac;
3h, X = SO$_3^-$, Y = H, Z = Ac; 4h. X = Y = H, Z = SO$_3^-$;
5h, X = Y = SO$_3^-$, Z = Ac; 6h, X = H, Y = Z = SO$_3^-$;
7h, X = Z = SO$_3^-$, Y = H; 8h, X = Y = Z = SO$_3^-$.
Ac = COCH$_3$

Heparin/Heparan Sulfate Disaccharides

c)

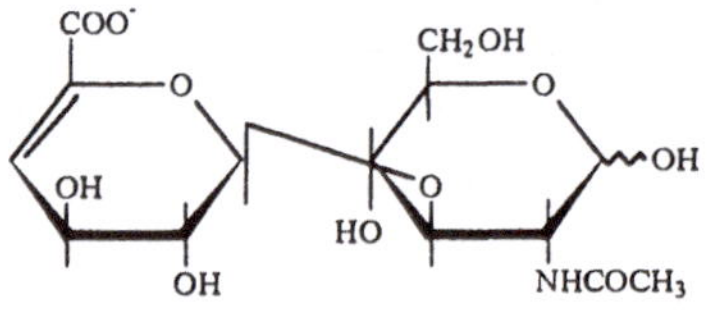

Hyaluronic Acid Disaccharide

Figure 5.30 Structures of the unsaturated disaccharide products formed enzymatically from (a) chondroitin sulfates/dermatan sulfates, (b) from heparin/ heparan sulfates and (c) from hyaluronic acid.

ment of chondroitin sulfate, dermatan sulfate and hyaluronic acid with polysaccharide lyases released 9 different disaccharides bearing unsaturated uronic acid residues that can be detected by UV absorbance at 232 nm without prior derivatization [131]. As illustrated in Figure 5.31, the disaccharides were separated on the basis of their net charges from –1 to –4, with the nonsulfated ones (peak 2) eluting first, followed by the mono- (peak 3–5) di- (peak 6–8) and trisulfated (peak 9) disaccharides (for structures see Figure 5.30a). A similar method was applied to separate eight commercial disaccharide standards prepared from heparin, heparan sulfate, and derivatized heparins by depolymerization with heparinases I, II and III [133]. Unlike in the former study two heparin/heparan sulfate disaccharides having an identical charge, were not fully resolved using a borate BGE. Resolution of these two disaccharides could be improved by either increasing the capillary length, preparing the buffer in deuterated water, or eliminating boric acid. Surprisingly, baseline resolution was achieved in SDS in the absence of buffer, although differential partitioning of the charged and polar disaccharides in the SDS micelles is unlikely. In a second step porcine mucosal heparin and bovine kidney heparan sulfate were depolymerized using heparinase, heparin lyase II and heparinitase and a combination of all three enzymes. CE analysis of the products formed provided the disaccharide composition of the two GAG's requiring as little as 15 ng of polysaccharide compared to 40 µg required for a comparable HPLC analysis.

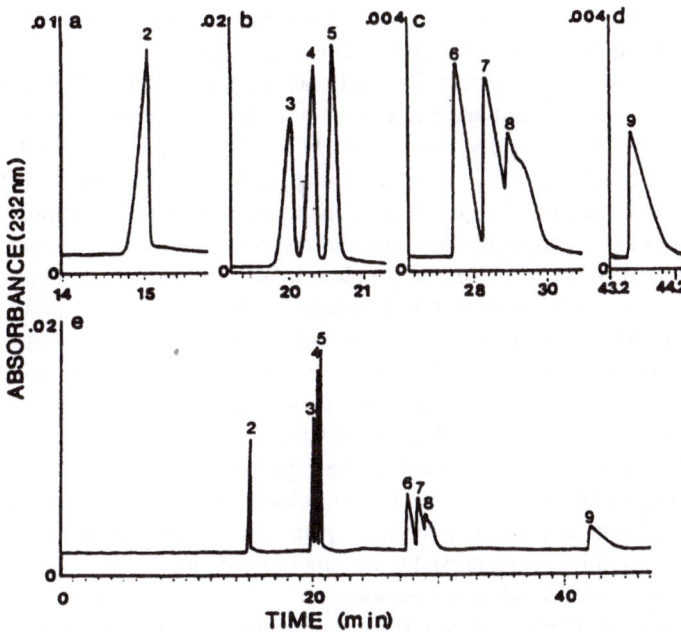

Figure 5.31 Separation of 1 nl of a mixture of chondroitin/dermatan sulfate derived disaccharides, containing 1 ng/nl each of the unsaturated disaccharides 2–9. (a–d) are expanded portions of electropherogram (e) showing the individual components 2, 3–5, 6–8, and 9, respectively. For peak assignment and structures see Figure 5.30a. Separation conditions: BGE, 10 mM sodium borate – 50 mM boric acid, pH 8.8; capillary, 68 cm x 75 μm ID; voltage, 10 kV; UV detection, 232 nm (from [131], with permission).

Carney and Osborne [132] further investigated the factors affecting the resolution of sulfated GAG disaccharides such as pH, borate concentration, ionic strength and the addition of SDS to the BGE. Although the disaccharides were too polar to partition with the SDS micelles, the presence of SDS in the BGE improved the resolution. The optimized separation conditions were applied to both nonsulfated hyaluronan hexa- to tetradecasaccharides and sulfated GAG disaccharides. The separation system was found to work well for the non- and monosulfated di- and oligosaccharides. However, the migration times for the di- and trisulfated disaccharides were to long resulting in poor peak shapes. Separation of these compounds in orthophosphoric acid under reversed polarity conditions improved the peak shape and the resolution tremendously. While non- and monosulfated disaccharides can be separated under alkaline conditions, more extensively sulfated species require acidic separation conditions and reversed polarity.

5.5.1 Heparins

Due to its vast structural heterogeneity unfractionated heparin displays a wide variety of biological effects not related to its anticoagulant or antithrombotic activity, which is a serious limitation in its value as a pharmaceutical drug. The production of heparin fragments with a more or less uniform mass distribution through gel filtration, depolymerization or chemical synthesis, relies very much on the availability of analytical procedures for the characterization of intermediates and final products. For that reason Damm et al. [134] developed CE methods for the analysis of natural and synthetic low molecular weight (LMW) heparin fragments. Using low pH electrolyte systems, namely 0.2 M phosphate pH 4.0, and controlling the capillary temperature at 40°C allowed the mapping of oligosaccharides from heparinase treated heparin and the quality assessment of synthetic pentasaccharide preparations. These pentasaccharides are identical or derivatives of the unique sequence responsible for the anticoagulant activity of heparin. A severe problem in the analysis of the synthetic pentasaccharide preparations occurred due to the detection mode applied. UV detection showed to be more sensitive for the synthetic precursors, which contain strong UV absorbing substituents, than for the end products, leading to an overestimation of the amount of contaminants. This problem was overcome by changing from the direct to the indirect UV detection mode, using sulfosalicylic acid or 1,2,4-tricarboxybenzoic acid as the BGE and the chromophore [135]. In contrast to direct UV detection, the signal obtained with indirect UV detection for the synthetic heparin pentasaccharides is almost independent of their molecular structure yielding representative peak areas and enabling quantitative analysis (Figure 5.32). Additionally, the sensitivity in indirect UV showed to be superior with a limit of detection of 5 femtomoles and a limit of quantitation of 25 femtomoles.

In a related study , Desai et al. [136] applied CE for the compositional analysis of heparin and different low molecular weight (LMW) heparin samples. Optimum separation of the oligosaccharides was achieved in a 10 mM borate buffer, pH 8.81, containing 50 mM SDS. Quantitative analysis could be performed with as little as 10 picomoles of an oligosaccharide sample. The oligosaccharide composition was obtained after almost complete depolymerization of the heparin samples with a mixture of heparin lyase I, II and III, with a mass balance of 95% for both heparin and LMW heparins. As determined by CE the oligosaccharide composition for the different LMW heparins varies indicating that their proportions of antithrombin III binding sequence differ significantly. CE of heparin and LMW heparins digested only with heparinase I revealed a high level of structural heterogeneity. The oligosaccharide maps thus obtained might be useful as a fingerprint for a given heparin sample allowing differentiation and quality control of commercially available LMW heparin preparations. Similar heparin derived oligosaccharides ranging in size from disaccharide to tetrasaccharide were separated using reverse polarity CE [137]. In an extensive comparative study the authors demonstrated that reverse polarity CE resolves disaccharide mixtures and mixtures of heparin derived oligosaccharides into all components using a single 20 mM phosphate buffer at pH 3.48. Comparison of the separation patterns obtained in the normal and the reverse polarity mode showed that the improved resolution in case of the latter one is mainly due to an increase in peak sharpness and improved peak symmetry, as illustrated in Figure 5.33. However, mixtures containing oligosaccharides primarily differing in size, such as oligosaccharides derived from enzymatic digest of bovine lung heparin, were better resolved under normal polarity conditions.

Beside natural and synthetic heparin fragments also heparinoid mimetics can be applied as anticoagulant and antithrombotic agents. A new class of heparinoid mimetics are spaced persulfated carbohydrates. Kinetic studies of these highly charged compounds in small volu-

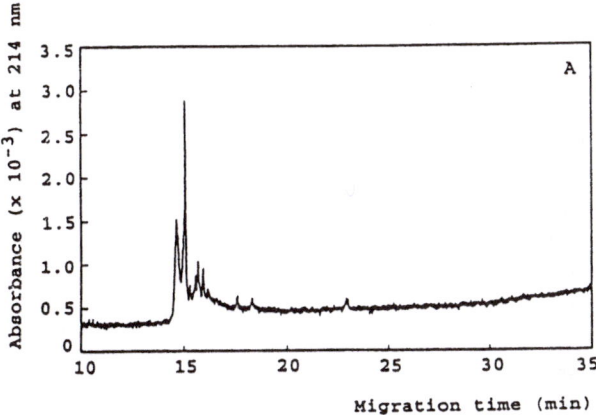

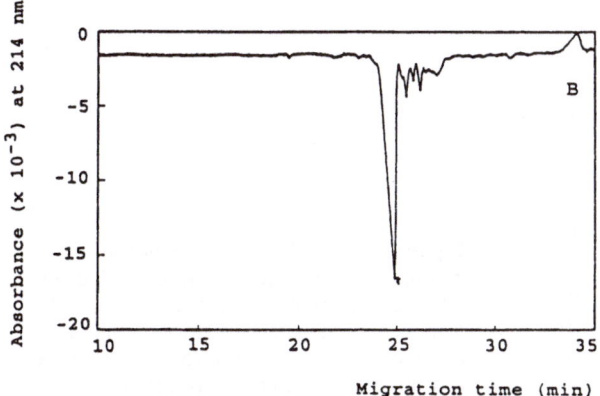

Figure 5.32 Analysis of pentasaccharide preparation HH2174 by CE using (A) direct and (B) indirect UV detection. Separation conditions: (A) BGE, 200 mM Na₂HPO₄, pH 3.0; capillary, 57 cm x 75 μm ID; voltage 7.5 kV; temperature, 40°C; UV detection, 210 nm; (B) BGE, 5 mM 5-sulfosalicylic acid, pH 3.0; capillary, 57 cm x 50 μm ID; voltage, 5 kV; temperature, 25°C; indirect UV detection, 214 nm. The concentration of the pentasaccharide in the sample solution is 1 mg/ml, the injection volume is (A) 50 nl or (B) 1.8 nl (from [135], with permission).

mes of rat plasma were performed using MECC with UV detection [138]. Plasma samples were introduced directly into the capillary after 1:1 dilution with 100 mM SDS. With a LOD of 3 μg/ml the sensitivity of the method proved to be sufficient to follow plasma levels in pharmacokinetic studies. The method showed good versatility for the separation of various heparinoid mimetics differing in aromatic groups or the number of sulfate groups. Another group of synthetic heparin-like pharmaceuticals are the sulfated *bis*-lactobionic acid amides. The problems associated with these species are that their synthesis usually yields molecules

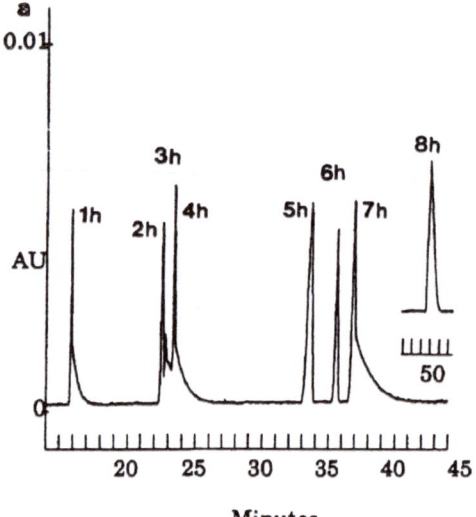

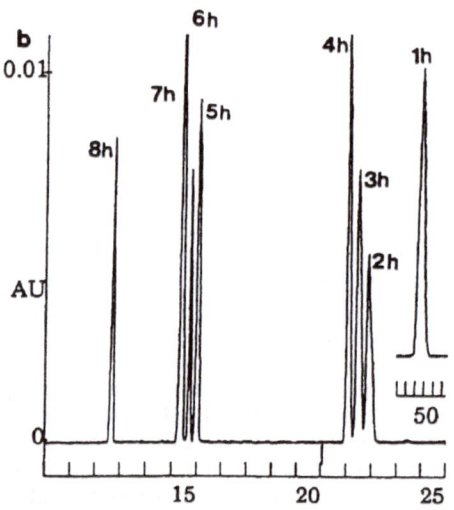

Figure 5.33 Capillary electrophoresis of a mixture of nonsulfated, monosulfated, disulfated, and trisulfated disaccharides derived from heparin/heparan sulfate. Separation conditions: capillary, 78 cm x 75 µm ID; UV detection, 232 nm; (a) BGE, 10 mM sodium borate – 50 mM SDS, pH 8.80; voltage, 12 kV (normal polarity); (b) BGE, 20 mM phosphoric acid, pH 3.48; voltage, –12 kV (reversed polarity). For peak assignment and structures see Figure 5.30b (from [137], with permission).

with differing degree of sulfation and that they may partially decompose during isolation. However, Malá et al. [139] demonstrated the suitability of CE and capillary isotachophoresis (CITP) for monitoring the composition of *bis*-lactobionic acid amide preparations. Taking advantage of the complex-forming equilibria between the analytes and bivalent metal cations present in the running electrolyte allowed for optimum separations.

The specificity of the enzymes acting on GAG's was investigated by treating homogeneous preparations of well defined heparin oligosaccharides with preparations of heparin lyase I, II and III [129]. The different sites of lyase action on the structurally defined oligosaccharides were examined by CE analysis of the produced oligosaccharide fragments. In a second step the heparin lyases were used to confirm the structure of an oligosaccharide based on the known specificity of the lyases.

5.5.2 Chondroitin and Dermatan

As already demonstrated by Carney and Osborne [132], the presence of borate in the BGE is advantageous in the separation of the chondroitin disaccharides. Separation of the three monosulfates Δ–2S, Δ–4S and Δ–6S (for structures see Figure 5.30a) was achieved using a borate buffer pH 8.8. Resolution of the Δ–4S and Δ–6S isomers was further improved using a mixed phosphate-borate buffer pH 9.0 containing SDS [132, 140, 141]. The addition of either triethylamine or cetyltrimethylammoium bromide to a borate containing BGE resulted in an improved separation due to an ion pairing effect [140]. Chondroitin derived disaccharides were also separated in an acidic phosphate BGE, pH 3.0, under reversed polarity conditions [142].

CE compositional analysis of GAG's was found to be suitable for the characterization of LMW dermatan sulfate as part of a structure-activity relationship study [143]. This study was undertaken to better understand the functioning of LMW dermatan sulfate as a non-heparin antithrombotic agent. Compositional analysis was accomplished by CE after extensive depolymerization of the LMW dermatan sulfate with chondroitin B or ABC lyase. This very general approach allowed also for the comparison of GAG's from different tissues and species.

A rapid, highly sensitive and reproducible CE method for the determination of all different types of unsaturated disaccharides present in pure or cell-secreted hyaluronan, chondroitin sulfate, dermatan sulfate and in proteoglycans was presented by Karamanos et al. [144]. Following chondroitinase digestion of the glycosaminoglycans or proteoglycans, the non-, di- and trisulfated Δ-disaccharides were completely resolved within 14 min. in a 15 mM phosphate BGE, pH 3.00, using reversed polarity. The proposed method was also applied for the sulfation analysis of tissue extracted proteoglycans. The sulfation patterns obtained were in close agreement with those obtained for purified GAG's. These results demonstrated the applicability of the present CE method for the analysis of biological preparations containing only minute amounts of GAG's which otherwise could be detected only by radiochemical techniques.

Denuzière et al. [145] applied CE of GAG derived disaccharides for the kinetic study of chondroitin sulfate digestion by the action of chondroitin ABC lyases. Although the main disaccharides formed upon depolymerization had identical charge and mass, the depolymerization process could be followed while maintaining baseline resolution of these isomers in a borate buffer. The method was applied to the comparison of the rate of hydrolysis of chondroitin sulfate with that of a complex associating chondroitin sulfate with chitosan. The experiments showed clearly that at physiological pH the chitosan protected the chondroitin sulfate from depolymerization.

5.5.3 Hyaluronan

Hyaluronic acid samples with different molecular masses (40'000–2'000'000) were quantitatively determined by CE in a bare fused silica capillary using a slightly acidic running buffer and UV detection at 185 nm [146]. The calibration curves showed good linearity from 0.01–3.3 mg/ml. Addition of pullunan to the running buffer induced a size exclusion mechanism leading to the migration of the hyaluronic acid samples according to their molecular mass. Hyaluronan could also be quantified in biological samples such as human and bovine vitreous [147]. Calibration standards were made up from known concentrations of hyaluronan from umbilical cord and the purity of the standard was examined by ^{1}H-NMR. Concentrations as low as 25 µg/ml could be detected. Identification of hyaluronan in the sample matrix was completed by depolymerization of the native mucopolysaccharide by hyaluronidase which resulted in the loss of the major hyaluronan peak and several new peaks with shorter migration times as indicated in Figure 5.34. When action of the hyaluronidase was allowed to proceed to completion the tetrasaccharide was the major product. In a mixed borate/phosphate BGE this tetrasaccharide migrated as a sharp peak and its area could be used to quantify the hyaluronan [148]. Due to the low intra- and inter-assay coefficient of variation highly reproducible results were obtained with this method. The monitoring of hyaluronan depolymerization products by CE was also adapted to the determination of the hyaluronidase activity in bee and snake venoms [149]. The crude source material was fractionated by HPLC prior to the enzymatic digestion.

Since patients with malignant mesothelioma often present effusions containing large amounts of hyaluronan, its quantitative determination can be used for diagnostic purposes. This was realized using CE, a 15 mM phosphate BGE pH 3.0, and reversed polarity [150]. The determination of hyaluronan in effusions was based on analysis of Δ4,5-disaccharides obtained by ethanol precipitation and subsequent digestion of the hyaluronan and present galactosaminoglycans with a mixture of chondroitinase ABC and AC. Heparan sulfate, proteins and glycoproteins also present in the effusion were separated by ultrafiltration and the hyaluronan-derived disaccharides were analyzed by direct injection of the filtrate into the CE system. To estimate the accuracy of the CE method, the hyaluronan levels in effusions from five healthy individuals and three patients with mesothelioma were determined with CE and HPLC. The results achieved with the two techniques were found to be in close agreement as indicated in Table 5.9.

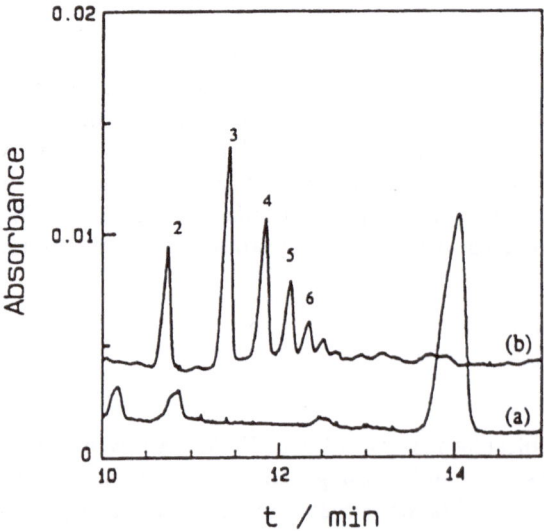

Figure 5.34 Electropherograms of (a) hyaluronan in bovine vitreous, compared with (b) hyaluronidase digest of hyaluronan. Separation conditions: BGE, 50 mM Na$_2$HPO$_4$ – 40 mM SDS – 10 mM Na$_2$B$_4$O$_7$; capillary, 50 cm x 75 µm ID; voltage, 15 kV (normal polarity); UV detection, 200 nm. Peaks are labeled according to the number of disaccharide units in the oligomer (from [147], with permission).

Table 5.9 Hyaluronan content in effusions determined by HPCE and HPLC[a] [150]

	HPCE	HPLC
Mesothelioma patients	45	46
	89	90
	323	318
Non-mesothelioma patients	1.6	1.7
	2.9	2.8
	9.3	9.3
	9.3	9.5
	16.8	18.6

[a] Results are expressed as the amount of hyaluronan derived GlcA in μg/ml effusion and are the average of three experiments in triplicate. Variations in the amounts of disaccharides measured were less than 5% in all cases.

5.5.4 Derivatized Glycosaminoglycans

Usually unsaturated GAG disaccharides are detected in the UV at 240 nm since they show a strong absorption at this wavelength, resulting in detection limits of about 40 ng/ml. [144]. Honda et al. [151] demonstrated for the first time the suitability of precolumn derivatization in combination with CE analysis of GAG disaccharides. Unsaturated disaccharides derived from GAG's by chondroitinase AC and ABC digestion were derivatized with PMP and separated in an alkaline borate BGE. The detection limit at 214 nm was at the 10 femtomole-level, calculated at the injected amount, with a linear range corresponding to 10–400 pg of injected amount. This system proved to be suitable for the determination of human urinary chondroitin sulfates. El Rassi et al. [152] selectively labeled GAG derived disaccharides via their carboxylic group with 7-aminonaphthalene-1,3-disulfonic acid (ANDSA). This labeling reaction comprises a condensation between the ANDSA amino group and the carboxylic group of the disaccharide in the presence of carbodiimide. The derivatization reaction yielded stable derivatives with percentage yields of about 97 % which could be detected by on-column LIF detection using a He-Cd-laser. With a LOD in the nanomolar range, sensitivity for the ANDSA derivatives was found to be three orders of magnitude higher compared to the detection of the underivatized disaccharides in the UV at 231 nm. Due to the presence of two sulfonic acid groups in the ANDSA tag, the disaccharides were easily separated at acidic pH, e.g. pH 4.0–5.0, using 100 mM sodium acetate as the BGE.

Another concept for the detection of carbohydrates is their *in situ* or dynamic labeling. Toida and Linhardt [153] recently reported the detection of GAG polymers as their copper (II) complexes. Except keratan sulfate which has no carboxylic groups at the complexation sites, all GAG polymers bind copper (II), forming a copper (II)-GAG complex with an absorption maximum close to 240 nm. Separation of the copper (II)-GAG complexes was performed under reversed polarity conditions at pH 4.7. The authors found the sensitivity of this method towards the GAG polymers to vary greatly with the structure of the substrate. Heparin showed the highest sensitivity with the sensitivity decreasing in the order: LMW heparin, dermatan sulfate, heparan sulfate and chondroitin sulfate. Hyaluronan also showed only very little sensitivity as its copper (II) complex. However, this method allowed the detection of nanogram amounts of copper (II)-heparin and the analysis of heparin oligosaccharides lacking unsaturated uronic acid residues due to their preparation.

5.6 Glycolipids

Another important class of glycoconjugates are the glycolipids among which the gangliosides are the most complex group. Gangliosides are found at high concentrations in the plasma membranes of vertebrate cells. They consist of a hydrophilic sialooligosaccharide chain and a hydrophobic ceramide moiety (see Figure 3.9). The many important biological functions and the lack of knowledge about their distribution and localization on cell surface membranes or in cytosolic compartments has generated a need for improved analytical methods.

Liu and Chan [154] demonstrated the suitability of CZE to study the micellar properties of gangliosides. As anionic amphiphilic compounds with hydrophilic head groups and hydrocarbon tails of similar sizes, gangliosides are known to form stable micelles in aqueous solutions. These micelles were separated within 10 min. with a mass sensitivity of 10^{-11} moles using uncoated fused-silica capillaries and UV detection at 195 nm. As shown in Figure 5.35, baseline resolution of a mixture of three ganglioside micelles, namely G_{M1}, G_{D1b} and G_{T1b} (for structures see Figure 3.9) was achieved in a 2.5 mM phosphate electrolyte at pH 7.40. The separation was mainly driven by the different degrees of sialylation of the three gangliosides. While the migration velocity of the ganglioside micelles was largely unaffected at pH 7.0–11.0, lowering the pH to 4.0 resulted in increasing migration times and peak broadening due to the diminished dissociation of carboxyl groups of the sialic acids [154]. The same effects were observed when increasing the ionic strength of the separation buffer. Liu and Chan also studied the time course of the formation of mixed ganglioside micelles. Shortly after mixing the G_{D1b} and the G_{T1b} micelles could be separated into two distinct peaks. Upon prolonged incubation at 37°C the two ganglioside peaks merged to form a single species within 2.5 h. This fusion process was found to be temperature dependent. While at 50°C formation of the mixed micelles was complete within 30 min., no fusion was observed at 0°C, even after 75 h. It was also demonstrated that the mixed micelle formation depends on the sialic acid content of the individual gangliosides.

In CE cyclodextrins (CD) are common buffer additives to alter selectivity and to improve resolution. Depending on the hydrophobicity and the size of the analytes CD's form inclusion complexes with a wide variety of compounds. Yoo et al. [155] demonstrated the suitability of CD's in the separation of the major gangliosides found in mammalian brains. Various parameters influencing the resolution such as the nature and the concentration of the CD and the pH of the BGE were investigated. Optimum separation regarding selectivity and efficiency was obtained using a 50 mM borate-phosphate BGE at pH 9.3, containing 16.5 mM α-CD. Detection was performed in the UV at 185 nm. It is assumed that the positive impact of α-CD on the ganglioside separation is due to the size of its cavity that best fits the lipid moiety. Applying these optimized separation conditions, several forms of gangliosides derived from extracts of deer antler, apricot seed and rat brain could be separated. While in deer antler G_{M1}, G_{D1a} and G_{D1b} were identified, apricot seeds and rat brain were found to contain G_{D1a}, and G_{M1} and G_{D1a}, respectively (Figure 5.36). With these results the authors could demonstrate the feasibility of CD modified CZE for detecting gangliosides in biological matrices which previously has been difficult due to the lack of chromophores and the mixed micelle formation.

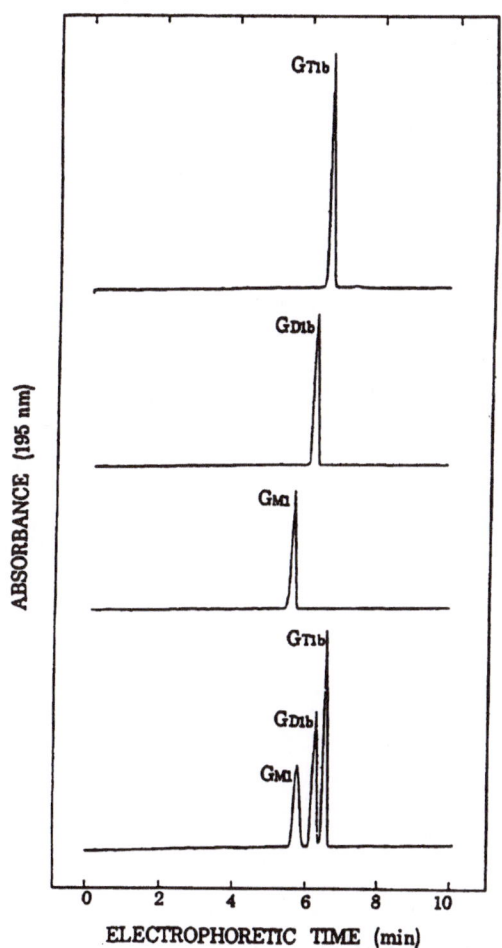

Figure 5.35 Capillary electrophoresis of the individual G_{M1}, G_{D1b}, and G_{T1b} micelles (10.7 µM) and a mixture of these three gangliosides shortly after mixing. Separation conditions: BGE, 2.5 mM potassium phosphate, pH 7.4; capillary, 72 cm x 50 µm ID; voltage, 30 kV; temperature, 30°C; UV detection, 195 nm (from [154], with permission).

A highly selective precolumn derivatization procedure for gangliosides was introduced, allowing for their sensitive UV and LIF detection [156]. This derivatization procedure which is based on the condensation of the carboxyl group of the sialic acid moiety with the amino group of ANDSA, has also been applied to the derivatization of GAG derived disaccharides (see Chapter 5.5). The strong sulfonic acid groups attached to the gangliosides through the ANDSA label allowed their separation over a wide pH range. Addition of acetonitrile or α-CD to the separation buffer resulted in a break-up of the amphiphilic ganglioside aggregates. Thus the gangliosides migrated as monomers in distinct peaks (Figure 5.37). A separation buffer containing 50 % acetonitrile in 5 mM sodium phosphate or 15 mM α-CD in 100 mM sodium borate was found to be ideal for the group separation of mono-, di- and trisialogangliosides with high resolution and efficiency [156]. Complete resolution of disialoganglioside isomers, such as the G_{D1a} and G_{D1b}, required the superimposition of a chromatographic component on the electrophoretic process by adding a hydrophilic (e.g. polyvinylalcohol, hydroxypropyl cellulose) or a hydrophobic selector (e.g. decanoyl-N-

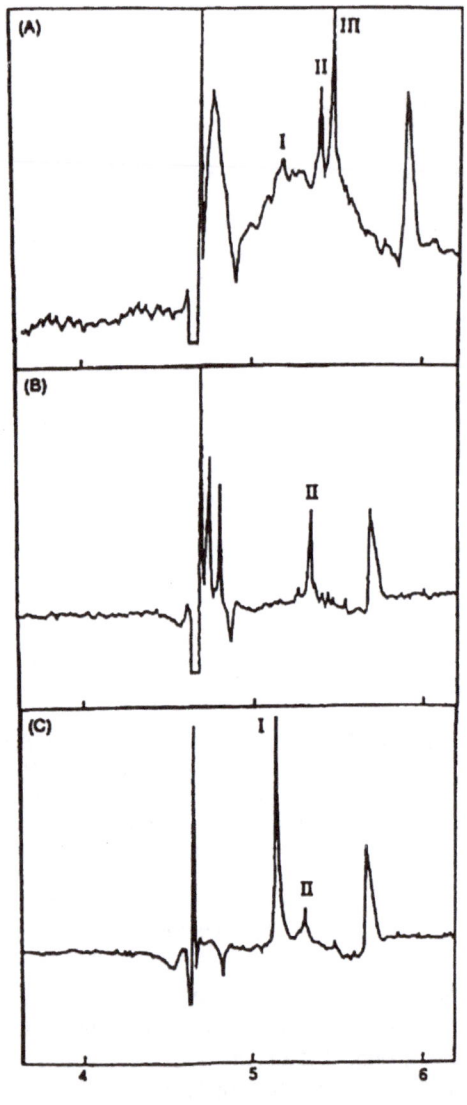

Figure 5.36 Electropherograms of gangliosides in biological matrices: (A) deer antler,
(B) apricot seed, (C) rat brain. Separation conditions: BGE, 20 mM borate – phosphate, pH
9.3, containing 16.5 mM α-CD; capillary, 60
cm x 50 µm ID; voltage, 30 kV; UV detection,
185 nm. Peak assignment: I = G_{M1}, II = G_{D1a},
III = G_{D1b} (from [155], with permission).

methylglucamide-borate surfactant complex) to the BGE. In a second step, Mechref et al.
[157] applied their novel derivatization procedure to sialooligosaccharides derived from
gangliosides by digestion with ceramideglycanase. The ceramideglycanase cleaved the gangliosides between the oligosaccharide chain and the ceramide of the glycosphingolipid resulting in a mixture of deconjugated oligosaccharides and ceramides. After isolation the
sialooligosaccharides were labeled with ANDSA and separated in an alkaline borate buffer.

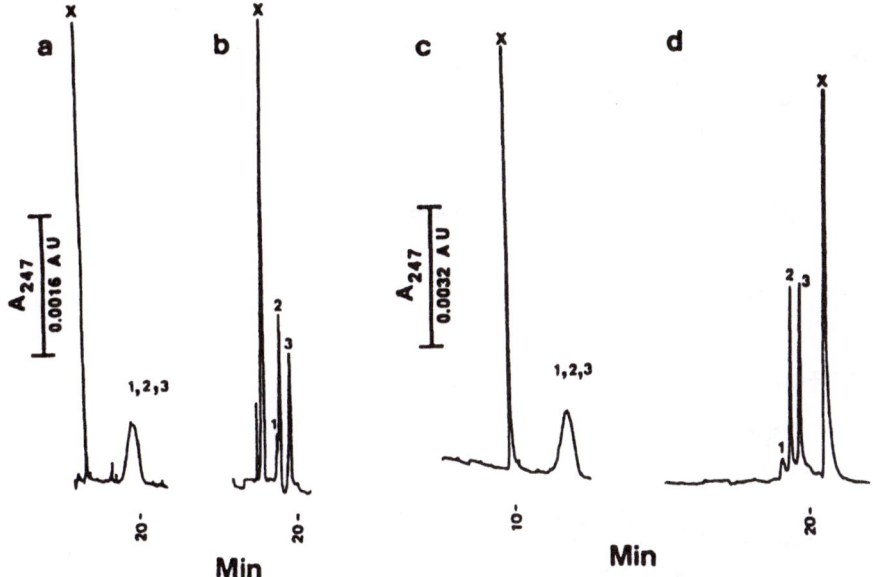

Figure 5.37 Electropherograms of standard gangliosides derivatized with ANDSA at neutral (a and b) and high pH (c and d) in the absence (a and c) and the presence (b and d) of ACN in the BGE. Separation conditions: BGE, in (a and b) 25 mM sodium phosphate, pH 7.0, at (a) 0% and (b) 50% (v/v) ACN; in (c and d) 10 mM sodium phosphate, pH 10.0, at (c) 0% and (d) 50% (v/v) ACN; capillary, 80/50 x 50 μm ID; voltage, in (a and b) 25 kV, in (c and d), 20 kV; UV detection, 247 nm. Peak assignment: $1 = G_{M1}$, $2 = G_{D1a}$, $3 = G_{T1b}$ (from [156], with permission).

5.7 Other glycoconjugates

Flavonoids are ubiquitous secondary plant metabolites which are widely used as remedies because of their spasmolytic, antiallergic and diuretic properties. In plants they are commonly present as flavonoid-O-glycosides in which one flavonoid hydroxyl group is bound to a sugar residue through an acid-labile hemiacetal bond. The 2-phenylbenzopyrone structure of flavonoids differ in the pattern of hydroxylation, the degree of unsaturation and the type and position of the sugar linkages (Figure 5.38). In two very extensive studies the suitability of CE for the separation of flavonoid-O-glycones differing in either their flavonoid aglycones [158] or in their sugar moiety [159] was demonstrated. The phenolic nature (pK_a = 10–12) of the flavonoids allows their dissociation at alkaline pH. Since the phenolic nature does not provide enough selectivity to allow sufficient resolution of the structurally closely related flavonoids, the authors investigated two other electrophoretic systems: one based on CZE with borate complexation and the other based on MECC. Using a neutral buffer containing SDS micelles a mixture of flavonoid-7-O-glycosides and their corresponding aglycones were resolved according to the hydrophobic interactions between the flavonoids and the hydrophobic core of the micelles [158]. Since the sugar residues render the flavonoids

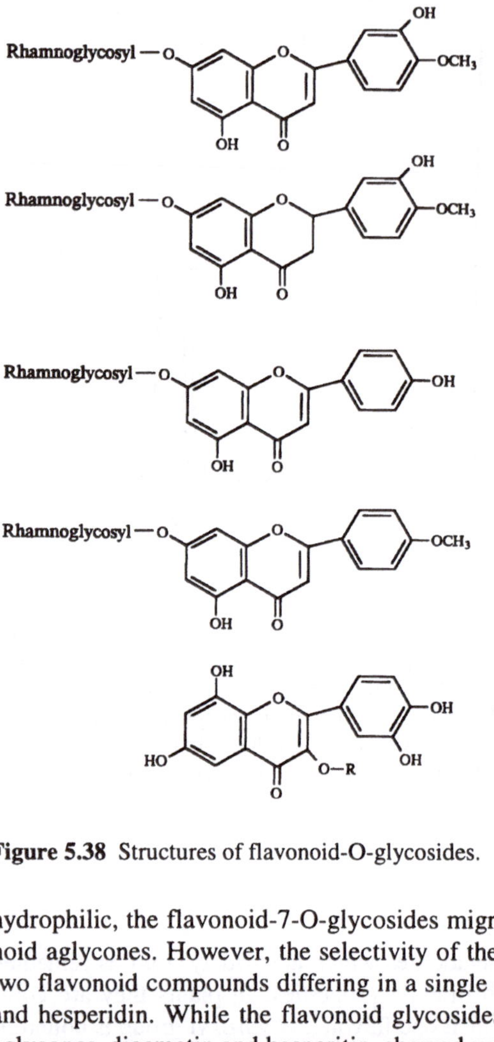

diosmin

hesperidin

isorhoifolin

linarin

quercetin-3-Oglycosides

R = α-L-Ara-β-D-Glc (1), β-D-Glc (2), β-D-Gal (3), α-L-Rha (4), α-L-Ara (5)

Figure 5.38 Structures of flavonoid-O-glycosides.

hydrophilic, the flavonoid-7-O-glycosides migrated faster than the more hydrophobic flavonoid aglycones. However, the selectivity of the MECC system was not sufficient to resolve two flavonoid compounds differing in a single carbon-carbon double bond, such as diosmin and hesperidin. While the flavonoid glycosides could not be resolved, their corresponding aglycones, diosmetin and hesperitin, showed very different migration times.

For the separation of a selected mixture of flavonoid-O-glycosides differing in their sugar moiety the use of CZE with borate complexation proved to be more useful [159]. In this case the selectivity is influenced mainly by *in situ* borate complexation of the sugar moieties and the cis-1,2-groups on the flavonoid skeleton. Ionization of the hydroxyl groups on the flavonoid skeleton due to the alkaline conditions had a rather slight impact on the separation. Figure 5.39 shows the separation of a mixture of quercetin-3-O-glycosides where the quercetin aglycone is associated with either a monosaccharide such as β-D-Glc, β-D-Gal, α-L-Rha or α-L-Ara, or with a disaccharide (α-L-Ara-β-D-Glc). For structures see Figure 5.38. As obvious from Figure 5.39, the use of an alkaline non-complexing buffer resulted only in partial resolution of the flavonoids, while the addition of the complexing borate ions allowed their complete separation. The same dependence of flavonoid resolution

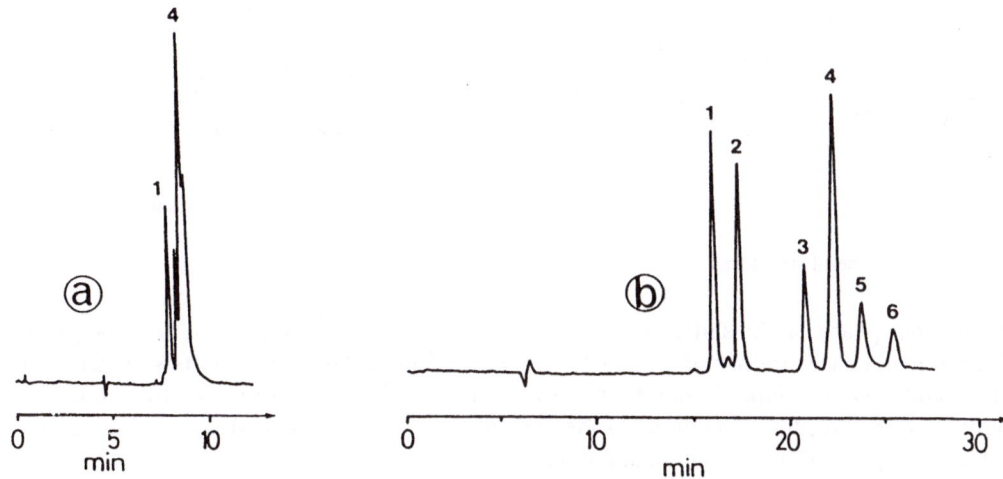

Figure 5.39 Influence of the complexing nature of the running buffer in free solution CE on the separation of flavonoid-3-O-glycosides. Separation conditions: BGE: in (a) 0.006 M Na$_2$B$_4$O$_7$ – 0.010 M NaH$_2$PO$_4$, pH 10.8, in (b) 0.2 M boric acid, pH 10.5; capillary, 65 cm x 50 µm ID; voltage, 24.4 kV; UV detection, 270 nm. Peaks 1–5 as in Figure 5.38, 6 = impurity (from [159], with permission).

on the borate concentration was found by Seitz et al. [160]. Using a 0.15 M borate buffer, pH 10, and an internal standard, they determined the content of rutin in a methanolic extract of *Sambuci flos*.

Borate complexation was also applied for the determination of flavonoids in sugar cane extracts [161] and for flavonols in fruit juices and wines [162]. Flavonols belong to the flavonoid group, having a higher degree of oxidation on the heterocycle. For the sugar cane flavonoids optimum separation was obtained using a 25 mM borate buffer, pH 9.5, containing 20% methanol. The addition of methanol was believed to induce changes in the solvation and stability of the flavonoid-borate complexes, resulting in altered migration order and improved selectivity [161]. The calibration curves based on peak area of a selected flavonoid showed a linear range from 10–200 pg/ml thus allowing accurate quantitation of flavonoids in sugar cane extract.

As an alternative to SDS as the micellar phase, cetyltrimethylammonium bromide (CTAB) and cholate in combination with taurine were investigated in MECC separations of flavonoids [163]. The CTAB based MECC system resulted in an extremely high selectivity due to both non-polar interactions with the aglycones and ion-pairing with the carboxylated flavonoids. In contrast, the cholate MECC system showed hydrophobic interaction with all flavonoids but to a lesser extent with the negatively charged ones due to electrostatic repulsion of the anionic micelles. Consequently, the different micellar systems induced differences in the selectivity and thus in the migration order of the flavonoids under investigation. Other MECC systems were developed for the separation and determination of cardiac glycosides from *Digitalis* species which are very important in the treatment of heart diseases [164]. Since these cardiac glycosides are very hydrophobic they migrate close to the migration time of the SDS micelles. However, the partition equilibrium between the micellar and

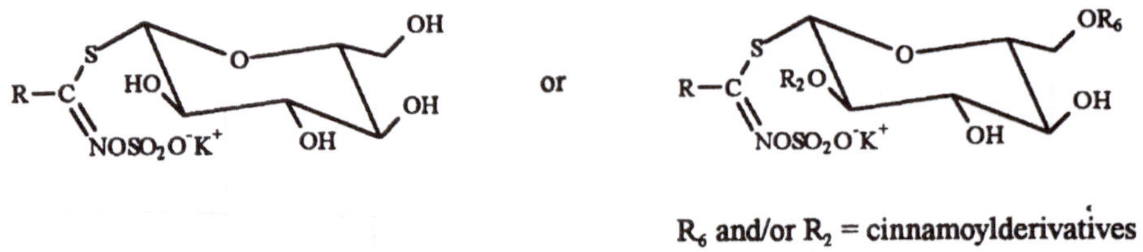

R₆ and/or R₂ = cinnamoylderivatives

R_6 and/or R_2 = cinnamoylderivatives

Figure 5.40 General structures of glucosinolates.

the aqueous phase can be shifted towards the aqueous phase trough addition of β- or γ-CD, through high concentrations of urea or the use of sodium cholate-borate buffers. The use of SDS and γ-CD containing borate buffers even allowed for the resolution of two anomeric cardiac glycosides, the glucodigifucoside and the glucodigiglucomethyloside, which could not be distinguished by conventional HPLC.

Glucosinolates are plant products with well defined structures as depicted in Figure 5.40. Of the more than 100 glucosinolates known only a few are quantitatively dominant in single plant species. Glucosinolates and their degradation products are important for the quality of food and feed based on glucosinolate containing plants such as oilseed rape, cabbage and kale, since high concentrations of these compounds were found to be antinutritive or toxic and to cause off-flavor effects. Michaelsen et al. [165] developed a MECC method using CTAB as the micellar phase for the qualitative and quantitative analysis of intact glucosinolates and desulfoglucosinolates. The separation was based on hydrophobic and ion-pairing interaction of the negatively charged glucosinolates with the positively charged micelles resulting in differential partitioning of the glucosinolates in the CTAB phase. The authors investigated the influence of applied voltage, buffer pH, the CTAB concentration and the buffer concentration on the separation performance, the reproducibility and the linearity. However, no single set of separation conditions was found to separate all glucosinolates under investigation.

References

1. A. Klockow, A. Paulus, V. Figueiredo, R. Amadò, and H.M. Widmer, "Determination of carbohydrates in fruit juices by capillary electrophoresis and high-performance liquid chromatography", *J. Chromatogr. A, 680* **1994** 187-200.

2. A.E. Vorndran, P.J. Oefner, H. Scherz, and G.K. Bonn, "Indirect UV detection of carbohydrates in capillary zone electrophoresis", *Chromatographia, 33* **1992** 163-168.

3. Y.-H. Lee, and T.-I. Lin, "Determination of carbohydrates by high-performance capillary electrophoresis with indirect absorbance detection", *J. Chromatogr. B, 681* **1996** 87-97.

4. A.M. Fermier, and L.A. Colon, "Capillary electrophoresis with constant potential amperometric detection using a nickel microelectrode for detection of carbohydrates", *J. High Resol. Chromatogr., 19* **1996** 613-616.

5. J. Ye, and R.P. Baldwin, "Determination of carbohydrates, sugar acids and alditols by capillary electrophoresis and electrochemical detection at a copper electrode", *J. Chromatogr. A, 687* **1994** 141-148.

6.   C.R. Noe, J. Freissmuth, D. Rothley, B. Lachmann, and P. Richter, "Kapillarelektrophoretische Analytik komplexer Kohlenhydratgemische", *Pharmazie, 51* **1996** 868-873.

7.   C. Huber, E. Grill, P.J. Oefner, and O. Bobleter, "Capillary electrophoretic determination of the component monosaccharides in hemicelluloses", *Fresenius J. Anal. Chem., 348* **1994** 825-831.

8.   C. Chiesa, P.J. Oefner, L.R. Zieske, and R.A. O'Neill, "Micellar electrokinetic chromatography of monosaccharides derivatized with 1-phenyl-3-methyl-2-pyrazolin-5-one", *J. Cap. Elec., 2* **1995** 175-183.

9.   J. Liu, O. Shirota, D. Wiesler, and M. Novotny, "Ultrasensitive fluorometric detection of carbohydrates as derivatives in mixtures separated by capillary electrophoresis", *Proc. Natl. Acad. Sci. USA, 88* **1991** 2302-2306.

10.   A. Guttman, "Analysis of monosaccharide composition by capillary electrophoresis", *J. Chromatogr. A, 763* **1997** 271-277.

11.   S. Honda, S. Iwase, A. Makino, and S. Fujiwara, "Simultaneous determination of reducing monosaccharides by capillary zone electrophoresis as the borate complexes of N-2-pyridylglycamines", *Anal. Biochem., 176* **1989** 72-77.

12.   T.J. O'Shea, S.M. Lunte, and W.R. LaCourse, "Detection of carbohydrates by capillary electrophoresis with pulsed amperometric detection", *Anal. Chem., 65* **1993** 948-951.

13.   R.E. Roberts, and D.C. Johnson, "Variation in PED Response at a gold microelectrode as a function of waveform parameters when applied to alditols and carbohydrates separated by capillary electrophoresis", *Electroanalysis, 7* **1995** 1015-1019.

14.   Z. Jin, R. Chen, and L.A. Colon, "Determination of glucose in submicroliter samples by CE-LIF using precolumn or on-column enzymatic reactions", *Anal. Chem., 69* **1997** 1326-1331.

15.   A.M. Arentoft, S. Michaelsen, and H. Sorensen, "Determination of oligosaccharides by capillary zone electrophoresis", *J. Chromatogr. A, 652* **1993** 517-524.

15a. T. Ogawa, T. Terabayashi, and Y. Kawanichi, "Analysis of oligo- and poly-N-acetylneuraminic acids and their lactones by capillary electrophoresis", *J. Chromatogr. A., 741* **1996** 295-298.

16.   W. Zhou, and R.P. Baldwin, "Capillary electrophoresis and electrochemical detection of underivatized oligo-and polysaccharides with surfactant controlled-electroosmotic flow", *Electrophoresis, 17* **1996** 319-324.

17.   W. Nashabeh, and Z. El Rassi, "Capillary zone electrophoresis of pyridylamino derivatives of maltooligosaccharides", *J. Chromatogr., 514* **1990** 57-64.

18.   S. Honda, S. Suzuki, A. Nose, K. Yamamoto, and K. Kakehi, "Capillary zone electrophoresis of reducing mono- and oligosaccharides as the borate complexes of their 3-methyl-1-phenyl-2-pyrazolin-5-one derivatives", *Carbohydr. Res., 215* **1991** 193-198.

19.   C. Chiesa, and C. Horváth, "Capillary zone electrophoresis of malto-oligosaccharides derivatized with 8-aminonaphthalene-1,3,6-trisulfonic acid", *J. Chromatogr., 645* **1993** 337-352.

20.   A. Klockow, H.M. Widmer, R. Amado, and A. Paulus, "Capillary electrophoresis of ANTS labelled oligosaccharide ladders and complex carbohydrates", *Fresenius J. Anal. Chem., 350* **1994** 415-425.

21.   M. Stefansson, and M.V. Novotny, "Separation of complex oligosaccharide mixtures by capillary electrophoresis in the open-tubular format", *Anal. Chem., 66* **1994** 1134-1140.

22.   M. Stefansson, and M.V. Novotny, "Resolution of the branched forms of oligosaccharides by high-performance capillary electrophoresis", *Carbohydr. Res., 258* **1994** 1-9.

23.   C. Chiesa, and R.A. O'Neill, "Capillary zone electrophoresis of oligosaccharides derivatized with various aminonaphthalene sulfonic acids", *Electrophoresis, 15* **1994** 1132-1140.

24.   J. Plocek, and M.V. Novotny, "Capillary zone electrophoresis of oligosaccharides derivatized with N-(4-aminobenzoyl)-L-glutamic acid for ultraviolet absorbance detection", *J. Chromatogr. A*, *757* **1997** 215-223.

25.   J. Sudor, and M.V. Novotny, "End-label, free-solution capillary electrophoresis of highly charged oligosaccharides", *Anal. Chem.*, *67* **1995** 4205-4209.

26.   A. Guttman, S. Brunet, and N. Cooke, "Capillary electrophoresis of carbohydrates in the biopharmaceutical and food and beverage industries", *LC-GC*, *14* **1996** 788-792.

27.   W. Nashabeh, and Z. El Rassi, "Capillary zone electrophoresis of linear and branched oligosaccharides", *J. Chromatogr.*, *600* **1992** 279-287.

28.   A. Rydlund, and O. Dahlman, "Oligosaccharides obtained by enzymatic hydrolysis of birch kraft pulp xylan: Analysis by capillary zone electrophoresis and mass spectrometry", *Carbohydr. Res.*, *300* **1997** 95-102.

29.   A. Rydlund, and O. Dahlman, "Rapid analysis of unsaturated acidic xylooligosaccharides from kraft pulps using CZE", *J. High Res. Chromatogr.*, *20* **1997** 72-76.

30.   A. Rydlund, and O. Dahlman, "Efficient capillary zone electrophoretic separation of wood-derived neutral and acidic mono- and disaccharides", *J. Chromatogr. A*, *738* **1996** 129-140.

31.   A. Nardi, S. Fanali, and F. Foret, "Capillary zone electrophoretic separation of cyclodextrins with indirect UV photometric detection", *Electrophoresis*, *11* **1990** 774-776.

32.   K. Lambertsen Larsen, F. Mathiesen, and W. Zimmermann, "Separation and analysis of cyclodextrins by capillary zone electrophoresis", *Carbohydr. Res.*, *298* **1997** 59-63.

33.   S.G. Penn, R.W. Chiu, and C.A. Monnig, "Separation and analysis of cyclodextrins by capillary electrophoresis with dynamic fluorescence labelling and detection", *J. Chromatogr. A*, *680* **1994** 233-241.

34.   T.-I. Lin, and Y.-H. Lee, "Capillary electrophoretic analysis of cyclodextrins and determination of formation constants for inclusion complexes", *Electrophoresis*, *17* **1996** 333-340.

35.   K.-B. Lee, Y.-S. Kim, and R.J. Linhardt, "Capillary zone electrophoresis for the quantitation of oligosaccharides formed through the action of chitinase", *Electrophoresis*, *12* **1991** 636-640.

36.   Z. Zhang, M.L. Pierce, and A.J. Mort, "Detection and differentiation of pectic enzyme activity in vitro and in vivo by capillary electrophoresis of products from fluorescent labeled substrate", *Electrophoresis*, *17* **1996** 372-378.

37.   A.J. Mort, and E.M.W. Chen, "Separation of 8-aminonaphthalene-1,3,6-trisulfonate (ANTS)-labeled oligomers containing galacturonic acid by capillary electrophoresis: Application to determining the substrate specifity of endopolygalacturonases", *Electrophoresis*, *17* **1996** 379-383.

38.   R. Zeleny, F. Altmann, and W. Praznik, "A capillary electrophoretic study on the specificity of β-galactosidases from Aspergillus oryzae, Escherichia coli, Streptococcus pneumoniae, and Canavalia ensiformis (Jack Bean)", *Anal. Biochem.*, *246* **1997** 96-101.

39.   X. Le, C. Scaman, Y. Zhang, J. Zhang, N.J. Dovichi, O. Hindsgaul, and M.M. Palcic, "Analysis by capillary electrophoresis-laser-induced fluorescence detection of oligosaccharides produced from enzyme reactions", *J. Chromatogr. A*, *716* **1995** 215-220.

40.   Y. Zhang, X. Le, N.J. Dovichi, C.A. Compston, M.M. Palcic, P. Diedrich, and O. Hindsgaul, "Monitoring biosynthetic transformations of N-acetyllactosamine using fluorescently labeled oligosaccharides and capillary electrophoretic separation", *Anal. Biochem.*, *227* **1995** 368-376.

41.   K.B. Lee, U.R. Desai, M.M. Palcic, O. Hindsgaul, and R.J. Linhardt, "An electrophoresis-based assay for glycosyltransferase activity", *Anal. Biochem.*, *205* **1992** 108-114.

42. X.C. Le, Y. Zhang, N.J. Dovichi, C.A. Compston, M.M. Palcic, R.J. Beever, and O. Hindsgaul, "Study of the enzymatic transformation of fluorescently labeled oligosaccharides in human epidermoid cells using capillary electrophoresis with laser-induced fluorescence detection", *J. Chromatogr. A*, *781* **1997** 515-522.

43. K.P. Bateman, J.H. Banoub, and P. Thibault, "Probing the microheterogeneity of O-specific chains from *Yersinia ruckeri* using capillary electrophoresis/electrospray mass spectrometry", *Electrophoresis*, *17* **1996** 1818-1828.

44. J. Kelly, H. Masoud, M.B. Perry, J.C. Richards, and P. Thibault, "Separation and characterization of O-deacetylated lipooligosaccharides and glycans derived from *Moraxella catarrhalis* using capillary electrophoresis-electrospray mass spectrometry and tandem mass spectrometry", *Anal. Biochem.*, *233* **1996** 15-30.

45. D.M. Richmond, and E.S. Yeung, "Development of laser-excited indirect fluorescence detection for high-molecular-weight polysaccharides in capillary electrophoresis", *Anal. Biochem.*, *210* **1993** 245-248.

46. J. Sudor, and M.V. Novotny, "Electromigration behavior of polysaccharides in capillary electrophoresis under pulsed-field conditions", *Proc. Natl. Acad. Sci. USA*, *90* **1993** 9451-9455.

47. H.-J. Zhong, M.A.K. Williams, R.D. Keenan, D.M. Goodall, and C. Rolin, "Separation and quantification of pectins using capillary electrophoresis: a prelaminary study", *Carbohydr. Polymers*, *32* **1997** 27-32.

48. M. Stefansson, and M.V. Novotny, "Modification of the electrophoretic mobility of neutral and charged polysaccharides", *Anal. Chem.*, *66* **1994** 3466-3471.

49. J.D. Brewster, and M.L. Fishman, "Capillary electrophoresis of plant starches as the iodine complexes", *J. Chromatogr. A*, *693* **1995** 382-387.

50. R.A. Dwek, C.J. Edge, D.J. Harvey, M.R. Wormald, and R.B. Parekh, "Analysis of glycoprotein-associated oligosaccharides", *Annu. Rev. Biochem.*, *62* **1993** 65-100.

51. N. Bihoreau, C. Ramon, M. Lazard, and J.M. Schmitter, "Combination of capillary electrophoresis and matrix-assisted laser desorption ionization mass spectrometry for glycosylation analysis of a human monoclonal anti-Rhesus (D) antibody", *J. Chromatogr. B*, *697* **1997** 123-133.

52. K. Kopp, M. Schlüter, and R.G. Werner, "Monitoring the glycosylation pattern of recombinant interferon-ω with high-pH anion-exchange chromatography and capillary electrophoresis", *Arzneim.-Forsch./Drug Res.*, *46* **1996** 1191-1196.

53. F. Kilár, and S. Hjertén, "Separation of the human transferrin isoforms by carrier-free high-performance zone electrophoresis and isoelectric focusing", *J. Chromatogr.*, *480* **1989** 351-357.

54. S.-L. Wu, G. Teshima, J. Cacial, and W.S. Hancock, "Use of high-performance capillary electrophoresis to monitor charge heterogeneity in recombinant-DNA derived proteins", *J. Chromatogr.*, *516* **1990** 115-122.

55. A.D. Tran, S. Park, P.J. Lisi, O.T. Huynh, R.R. Ryall, and P.A. Lane, "Separation of carbohydrate-mediated microheterogeneity of recombinant human erythropoietin by free solution capillary electrophoresis", *J. Chromatogr.*, *542* **1991** 459-471.

56. E. Watson, and F. Yao, "Capillary electrophoretic separation of human recombinant erythropoietin (r-HuEPO) glycoforms", *Anal. Biochem.*, *210* **1993** 389-393.

57. H.P. Bietlot, and M. Girard, "Analysis of recombinant human erythropoietin in drug formulations by high-performance capillary electrophoresis", *J. Chromatogr. A*, *759* **1997** 177-184.

58.   P.D. Grossman, J.C. Colburn, H. Lauer, R.G. Nielsen, R.M. Riggin, G.S. Sittampalan, and E.C. Rickard, "Application of free-solution capillary electrophoresis to the analytical scale separation of proteins and peptides", *Anal. Chem., 61* **1989** 1186-1194.

59.   P.M. Rudd, I.G. Scragg, E. Coghill, and R.A. Dwek, "Separation and analysis of the glycoform populations of ribonuclease B using capillary electrophoresis", *Glycoconjugate J., 9* **1992** 86-91.

60.   S. Honda, A. Makino, S. Suzuki, and K. Kakehi, "Analysis of oligosaccharides in ovalbumin by high-performance capillary electrophoresis", *Anal. Biochem., 191* **1990** 228-234.

61.   P.J. Landers, R.P. Oda, B.J. Madden, and T.C. Spelsberg, "High-performance capillary electrophoresis of glycoproteins: The use of modifiers of electroosmotic flow for analysis of microheterogeneity", *Anal. Biochem., 205* **1992** 115-124.

62.   R.P. Oda, B.J. Madden, T.C. Spelsberg, and J.P. Landers, "α, ω-Bis-quarternary ammonium alkanes as effective buffer additives for enhanced capillary electrophoretic separation of glycoproteins", *J. Chromatogr. A, 680* **1994** 85-92.

63.   M.E. Legaz, and M.M. Pedrosa, "Effect of polyamines on the separation of ovalbumin glycoforms by capillary electrophoresis", *J. Chromatogr. A, 719* **1996** 159-170.

64.   D.E. Morbeck, B.J. Madden, and D.J. McCormick, "Analysis of the microheterogeneity of the glycoprotein chorionic gonadotropin with high-performance capillary electrophoresis", *J. Chromatogr. A, 680* **1994** 217-224.

65.   P. Laidler, D.A. Cowan, R.C. Hider, and A.T. Kicman, "Characterization of human chorionic gonadotropin microheterogeneity by capillary electrophoresis: potential application for quality control in the pharmaceutical industry", *Pharm. Sci., 3* **1997** 487-491.

66.   K.W. Yim, "Fractionation of the recombinant tissue plasminogen activator (rtPA) glycoforms by high-performance capillary electrophoresis and capillary isoelectric focusing", *J. Chromatogr. A, 559* **1991** 401-410.

67.   J.M. Thorne, W.K. Goetzinger, A.B. Chen, K.G. Moorhouse, and B.L. Karger, "Examination of capillary zone electrophoresis, isoelectric focusing and sodium dodecyl sulfate capillary electrophoresis for the analysis of recombinant tissue plasminogen activator", *J. Chromatogr. A, 744* **1996** 155-165.

68.   A.B. Chen, C.A. Rickel, A. Flanigan, G. Hunt, and K.G. Moorhouse, "Comparison of ampholytes used for slab gel and capillary isoelectric focusing of recombinant tissue-type plasminogen activator glycoforms", *J. Chromatogr. A, 744* **1996** 279-284.

69.   K.G. Moorhouse, C.A. Eusebio, G. Hunt, and A.B. Chen, "Rapid one-step capillary isoelectric focusing method to monitor charged glycoforms of recombinant human tissue-type plasminogen activator", *J. Chromatogr. A, 717* **1995** 61-69.

70.   K.G. Moorhouse, C.A. Rickel, and A.B. Chen, "Electrophoretic separation of recombinant tissue-type plasminogen activator glycoforms: Validation issues for capillary isoelectric focusing methods", *Electrophoresis, 17* **1996** 423-430.

71.   F. Kilár, and S. Hjertén, "Unfolding of human serum transferrin in urea studied by high-performance capillary electrophoresis", *J. Chromatogr., 638* **1993** 269-276.

72.   R.P. Oda, and J.P. Landers, "Effect of cationic buffer additives on the capillary electrophoretic separation of serum transferrin from different species", *Electrophoresis, 17* **1996** 431-437.

73.   R.P. Oda, R. Prasad, R.L. Stout, D. Coffin, W.P. Patton, D.L. Kraft, J.F. O'Brien, and J.P. Landers, "Capillary electrophoresis-based separation of transferrin sialoforms in patients with carbohydrate-deficient glycoprotein syndrome", *Electrophoresis, 18* **1997** 1819-1826.

74.   R. Prasad, R. Stout, D. Coffin, and J. Smith, "Analysis of carbohydrate deficient transferrin by capillary zone electrophoresis", *Electrophoresis, 17* **1997** 1814-1818.

75. D.C. James, R.B. Freedman, M. Hoare, and N. Jenkins, "High-resolution separation of recombinant human interferon-γ glycoforms by micellar electrokinetic capillary chromatography", *Anal. Biochem., 222* **1994** 315-322.

76. R. Vincentelli, and N. Bihoreau, "Characterization of each isoform of a F(ab')₂ by capillary electrophoresis", *J. Chromatogr., 641* **1993** 383-390.

77. S. Hoffstetter-Kuhn, G. Alt, and R. Kuhn, "Profiling of oligosaccharide-mediated microheterogeneity of a monoclonal antibody by capillary electrophoresis", *Electrophoresis, 17* **1996** 418-422.

78. A. Rice, J. Grimshaw, J. Trocha-Grimshaw, P. McCarron, and G.B. Wisdom, "Identification of α_1-acid glycoprotein (orosomucoid) in human synovial fluid by capillary electrophoresis", *J. Chromatogr. A, 722* **1997** 305-311.

79. J. Pedersen, and K. Biederman, "Characterization of proteinase A glycoforms from recombinant *Saccharomyces cerevisiae*", *Biotechnol. Appl. Biochem., 18* **1993** 377-388.

80. K. Yim, J. Abrams, and A. Hsu, "Capillary zone electrophoretic resolution of recombinant human bone morphogenetic protein 2 glycoforms. An investigation into the separation mechanisms for an exquisite separation", *J. Chromatogr. A, 716* **1995** 401-412.

81. N.K. Klausen, and T. Kornfelt, "Analysis of the glycoforms of human recombinant factor VIIa by capillary electrophoresis and high-performance liquid chromatography", *J. Chromatogr. A, 718* **1995** 195-202.

82. E. Watson, and F. Yao, "Capillary electrophoretic separation of recombinant granulocyte-colony-stimulating factor glycoforms", *J. Chromatogr., 630* **1993** 442-446.

83. M.J. Schmerr, and K.R. Goodwin, "Characterization by capillary electrophoresis of the surface glycoproteins of ovine lentiviruses before and after treatment with glycosidic enzymes", *J. Chromatogr. A, 652* **1993** 199-205.

84. M.G. Trevino, M. Taverna, H. Boureau, P. Bourlioux, and D. Ferrier, "Contribution of capillary zone electrophoresis to the analysis of cecal mucins", *J. Cap. Elec., 3* **1996** 287-294.

85. W.E. Werner, D.M. Demorest, and J.E. Wiktorowicz, "Automated Ferguson analysis of glycoproteins by capillary electrophoresis using a replaceable sieving matrix", *Electrophoresis, 14* **1993** 759-763.

86. R. Bonifichi, "Capillary electrophoresis of glycosylated proteins performed on a conventional capillary gas chromatographic column", *J. Chromatogr. A, 741* **1996** 139-145.

87. M. Huang, J. Plocek, and M.V. Novotny, "Hydrolytically stable cellulose-derivative coatings for capillary electrophoresis of peptides, proteins and glycoconjugates", *Electrophoresis, 16* **1995** 396-401.

88. Z. El Rassi (ed.), in *"Carbohydrate Analysis"*, Journal of Chromatography Library, Vol. 58, Elsevier, **1995**, Amsterdam.

89. J.F. Kelly, S.J. Locke, L. Ramaley, and P. Thibault, "Development of electrophoretic conditions for the characterization of protein glycoforms by capillary electrophoresis-electrospray mass spectrometry", *J. Chromatogr. A, 720* **1996** 409-427.

90. B. Yeung, T.J. Porter, and J.E. Vath, "Direct isoform analysis of high-mannose-containing glycoproteins by on-line capillary electrophoresis electrospray mass spectrometry", *Anal. Chem., 69* **1997** 2510-2516.

91. L. Yang, Q. Tang, A.K. Harrata, and C.S. Lee, "Capillary isoelectric focusing-electrospray ionization mass spectrometry for transferrin glycoform analysis", *Anal. Biochem., 243* **1996** 140-149.

92. J.A. Chakel, E. Pungor, W.S. Hancock, and S.A. Swedberg, "Analysis of DNA-derived glycoproteins via high-perfomance capillary electrophoresis coupled with off-line matrix-assisted laser desorption ionization time-of-flight mass spectrometry", *J. Chromatogr. B, 689* **1997** 215-220.

93. W. Nashabeh, and Z. El Rassi, "Capillary zone electrophoresis of α_1-acid glycoprotein fragments from trypsin and endoglycosidase digestions", *J. Chromatogr., 536* **1991** 31-42.

94. M. Taverna, A. Baillet, D. Biou, M. Schlüter, R. Werner, and D. Ferrier, "Analysis of carbohydrate-mediated heterogeneity and characterization of N-linked oligosaccharides of glycoproteins by high performance capillary electrophoresis", *Electrophoresis, 13* **1992** 359-366.

95. J. Jiskra, V. Pacakova, M. Ticha, K. Stulik, and T. Barth, "Use of capillary electrophoresis and high-performance liquid chromatography for monitoring of glycosylation of the peptides dalargin and desmopressin", *J. Chromatogr. A, 761* **1997** 285-296.

96. S.R. Rush, P.L. Derby, T.W. Strickland, and M.F. Rohde, "Peptide mapping and evaluation of glycopeptide microheterogeneity derived from endoproteinase digestion of erythropoietin by affinity high-performance capillary electrophoresis", *Anal. Chem., 65* **1993** 1834-1842.

97. D.L. LeTourneau, and N.E. Allen, "Use of capillary electrophoresis to measure dimerization of glycopeptide antibiotics", *Anal. Biochem., 246* **1997** 62-66.

98. P.L. Weber, T. Kornfelt, N.K. Klausen, and S.M. Lunte, "Characterization of glycopeptides from recombinant coagulation factor VIIa by high-performance liquid chromatography and capillary zone electrophoresis using ultraviolet and pulsed electrochemical detection", *Anal. Biochem., 225* **1995** 135-142.

99. P.L. Weber, and S.M. Lunte, "Capillary electrophoresis with pulsed amperometric detection of carbohydrates and glycopeptides", *Electrophoresis, 17* **1996** 302-309.

100. A.P. Hunter, and D.E. Games, "Evaluation of glycosylation site heterogeneity and selective identification of glycopeptides in proteolytic digests of bovine α_1-acid glycoprotein by mass spectrometry", *Rapid Communication in Mass Spectrometry, 9* **1995** 42-56.

101. S.-L. Wu, "The use of sequential high-performance liquid chromatography and capillary zone electrophoresis to separate the glycosylated peptides from recombinant tissue plasminogen activator to a detailed level of microheterogeneity", *Anal. Biochem., 253* **1997** 85-97.

102. K.R. Birdwell, T.L. Austell, R.S. Black, J.W. Jorgenson, and R.G. Hiskey, "Evaluation of proteolytically released carbohydrate-containing peptides of bovine prothrombin fragment 1 using electrospray ionization mass spectrometry and capillary electrophoresis", *J. Liq. Chrom. & Rel. Technol., 20* **1997** 987-1004.

103. A. Klockow-Beck, and A. Paulus, "Carbohydrate analysis with capillary electrophoresis", in *"A laboratory guide to glycoconjugate analysis"* (ed. P. Jackson and J.T. Gallagher), Birkhäuser, **1997**, Basel,141-158.

104. O'Neill, "Enzymatic release of oligosaccharides from glycoproteins for chromatographic and electrophoretic analysis", *J. Chromatogr. A, 720* **1996** 201-215.

105. S. Suzuki, K. Kakehi, and S. Honda, "Two-dimensional mapping of N-glycosidically linked asialo-oligosaccharides from glycoproteins as reductively pyridylaminated derivatives using dual separation modes of high-performance capillary electrophoresis", *Anal. Biochem., 205* **1992** 227-236.

106. S.W. Yuen, L.R. Zieske, I.M. Zaidi, D. Fu, and R.A. O'Neill, "A facile method for the release, labeling and CE analysis of glycoprotein oligosaccharides", *Techniques in Protein Chemistry, 5* **1994** 275-284.

107. P. Hermentin, R. Witzel, R. Doenges, R. Bauer, H. Haupt, T. Patel, R.B. Parekh, and D. Brazel, "The mapping by high-pH anion-exchange chromatography with pulsed amperometric detec

tion and capillary electrophoresis of the carbohydrate moieties of human plasma α_1-acid gly-coprotein", *Anal. Biochem., 206* **1992** 419-429.

108. P. Hermentin, R. Doenges, R. Witzel, C.H. Hokke, J.F.G. Vliegenhart, P. Kamerling, H.S. Conradt, M. Nimtz, and D. Brazel, "A strategy for the mapping of N-glycans by high-performance capillary electrophoresis", *Anal. Biochem., 221* **1994** 29-41.

109. M. Taverna, A. Baillet, and D. Baylocq-Ferrier, "Analysis of neutral and sialylated N-linked oligosaccharides by micellar electrokinetic capillary chromatography with addition of a diva-lent cation", *Chromatographia, 37* **1993** 415-422.

110. M. Taverna, A. Baillet, M. Schlüter, and D. Baylocq-Ferrier, "N-Glycosylation site mapping of recombinant tissue plasminogen activator by micellar electrokinetic capillary chromatogra-phy", *Biomed. Chromatogr., 9* **1995** 59-67.

111. K.F. Greve, D.E. Hughes, and B.L. Karger, "Capillary electrophoretic examination of un-derivatized mixtures released from immunoglobulin G antibodies and CTLA4Ig fusion pro-tein", *J. Chromatogr. A, 749* **1996** 237-245.

112. K. Kakehi, A. Susami, A. Taga, S. Suzuki, and S. Honda, "High-performance capillary electro-phoresis of O-glycosidically linked sialic acid-containing oligosaccharides in glycoproteins as their alditol derivatives with low-wavelength UV monitoring", *J. Chromatogr. A, 680* **1994** 209-215.

113. P. Jackson, "The use of polyacrylamide-gel electrophoresis for the high-resolution separation of reducing saccharides labelled with the fluorophore 8-aminonaphthalene-1,3,6-trisulphonic acid", *Biochem. J., 270* **1990** 705-713.

114. P. Jackson, "Polyacrylamide gel electrophoresis of reducing saccharides labeled with the fluo-rophore 2-aminoacridone: subpicomolar detection using an imaging system based on a cooled charge-coupled device", *Anal. Biochem., 196* **1991** 238-244.

115. A. Klockow, R. Amado, H.M. Widmer, and A. Paulus, "Separation of 8-aminonaphthalene-1,3,6-trisulfonic acid -labelled neutral and sialylated N-linked complex oligosaccharides by capillary electrophoresis", *J. Chromatogr. A, 716* **1995** 241-257.

116. P. Camilleri, G.B. Harland, and G. Okafo, "High resolution and rapid analysis of branched oligosaccharides by capillary electrophoresis", *Anal. Biochem., 230* **1995** 115-122.

117. A. Paulus, and A. Klockow, "Detection of carbohydrates in capillary electrophoresis", *J. Chromatogr. A, 720* **1996** 353-376.

118. A. Guttman, and C. Starr, "Capillary and slab gel electrophoresis profiling of oligosaccharides", *Electrophoresis, 16* **1995** 993-997.

119. G.B. Harland, G. Okafo, P. Matejtschuk, I.C. Sellick, G.E. Chapman, and P. Camilleri, "Fin-gerprinting of glycans as their 2-aminoacridone derivatives by capillary electrophoresis and laser-induced fluorescence", *Electrophoresis, 17* **1996** 406-411.

120. G. Okafo, L.M. Burrow, W. Neville, A. Truneh, R.A.G. Smith, M. Reff, and P. Camilleri, "Simple differentiation between core-fucosylated and nonfucosylated glycans by capillary electrophoresis", *Anal. Biochem., 240* **1996** 68-74.

121. G. Okafo, L. Burrow, S.A. Cary, G.D. Roberts, W. Johnson, and P. Camilleri, "A coordinated high-performance liquid chromatographic, capillary electrophoretic, and mass spectrometric approach for the analysis of oligosaccharide mixtures derivatized with 2-aminoacridone", *Anal. Chem., 68* **1996** 4424-4430.

122. A. Guttman, and T. Pritchett, "Capillary gel electrophoresis separation of high-mannose type oligosaccharides derivatized by 1-aminopyrene-3,6,8-trisulfonic acid", *Electrophoresis, 16* **1995** 1906-1911.

123. A. Guttman, F.-T.A. Chen, and R.A. Evangelista, "Separation of 1-aminopyrene-3,6,8-trisulfonate-labeled asparagine-linked fetuin glycans by capillary gel electrophoresis", *Electrophoresis, 17* **1996** 412-417.

124. A. Guttman, and S. Herrick, "Effect of the quantity and linkage position of mannose (α1,2) residues in capillary gel electrophoresis of high mannose type oligosaccharides", *Anal. Biochem., 235* **1996** 236-239.

125. A. Guttman, and K. Williams Ulfelder, "Exoglycosidase matrix-mediated sequencing of a complex glycan pool by capillary electrophoresis", *J. Chromatogr. A, 781* **1997** 547-554.

126. A. Guttman, "Multistructure sequencing of N-linked fetuin glycans by capillary gel electrophoresis and enzyme matrix digestion", *Electrophoresis, 18* **1997** 1136-1141.

127. H. Suzuki, O. Müller, A. Guttman, and B.L. Karger, "Analysis of 1-aminopyrene-3,6,8-trisulfonate derivatized oligosaccharides by capillary electrophoresis with matrix-assisted laser desorption/ionization time-of-flight mass spectrometry", *Anal. Chem., 69* **1997** 4554-4559.

128. K.N. Gu, R.J. Linhardt, M. Laliberte, K. Gu, and J. Zimmermann, "Purification, characterization and specificity of chondroitin lyases and glycuronidase from *Flavobacterium heparinum*", *Biochem. J., 312* **1995** 569-577.

129. U.R. Desai, H.-M. Wang, and R.J. Linhardt, "Substrate specifity of the heparin lyases from *flavobacterium heparinum*", *Arch. Biochem. Biophys., 306* **1993** 461-468.

130. J. Grimshaw, "Analysis of gycosaminoglycans and their oligosaccharide fragments by capillary electrophoresis", *Electrophoresis, 18* **1997** 2408-2414.

131. A. Al-Hakim, and R.J. Linhardt, "Capillary electrophoresis for the analysis of chondroitin sulfate- and dermatan sulfate-derived disaccharides", *Anal. Biochem., 195* **1991** 68-73.

132. S.L. Carney, and D.J. Osborne, "The separation of chondroitin sulfate disaccharides and hyaluronan oligosaccharides by capillary zone electrophoresis", *Anal. Biochem., 195* **1991** 132-140.

133. S.A. Ampofo, H.-M. Wang, and R.J. Linhardt, "Disaccharide compositional analysis of heparin and heparan sulfate using capillary electrophoresis", *Anal. Biochem., 199* **1991** 249-255.

134. J.B.L. Damm, G.T. Overklift, B.M.W. Vermeulen, C.F. Fluitsma, and G.W.K. van Dedem, "Separation of natural and synthetic heparin fragments by high-performance capillary electrophoresis", *J. Chromatogr., 608* **1992** 297-309.

135. J.B.L. Damm, and G.T. Overklift, "Indirect UV detection as a non-selective detection method in the qualitative and quantitative analysis of heparin fragments by high-performance capillary electrophoresis", *J. Chromatogr. A, 678* **1994** 151-165.

136. U.R. Desai, H.-M. Wang, S.A. Ampofo, and R.J. Linhardt, "Oligosaccharide composition of heparin and low-molecular weight heparins by capillary electrophoresis", *Anal. Biochem., 213* **1993** 120-127.

137. A. Pervin, A. Al-Hakim, and R.J.Linhardt, "Separation of glycosaminoglycan-derived oligosaccharides by capillary electrophoresis using reversed polarity", *Anal. Biochem., 221* **1994** 182-188.

138. S. Mayer, and M. Schleimer, "Quantitative determination of heparinoid mimetics in human and rat plasma by micellar electrokinetic chromatography", *J. Chromatogr. A, 730* **1996** 297-303.

139. Z. Mala, L. Krivankova, and P. Bocek, "Analysis of heparin-like pharmaceuticals by capillary zone electrophoresis and isotachophoresis", *Electrophoresis, 17* **1996** 125-129.

140. L. Scapol, E. Marchi, and G.C. Viscomi, "Capillary electrophoresis of heparin and dermatan sulfate unsaturated disaccharides with triethylamine and acetonitrile as electrolyte additives", *J. Chromatogr. A, 735* **1996** 367-374.

141. S. Michaelsen, M. Schroder, and H. Sorensen, "Separation and determination of glycosamino-glycan disaccharides by micellar electrokinetic chromatography for studies of pelt glycosami-noglycans", *J. Chromatogr. A, 652* **1993** 503-515.

142. E.P. Lillehoj, and S.S. Alexander, "Viring associated trans-regulatory protein of human leuke-mia virus type I", *Aids Res. Human Retrovir., 8* **1992** 237-244.

143. R.J. Linhardt, U.R. Desai, J. Liu, A. Pervin, D. Hoppenstaedt, and J. Fareed, "Low molecular weight dermatan sulfate as an antithrombotic agent", *Biochem. Pharmacol., 47* **1994** 1241-1252.

144. N.K. Karamanos, S. Axelsson, P. Vanky, G.N. Tzanakakis, and A. Hjerpe, "Determination of hyaluronan and galactosaminoglycan disaccharides by high-performance capillary electropho-resis at the attomole level. Applications to analysis of tissue and cell culture proteoglycans", *J. Chromatogr. A, 696* **1995** 295-305.

145. A. Denuziere, M. Taverna, D. Ferrier, and A. Domard, "Capillary electrophoresis of glycosami-noglycan-derived disaccharides: Application to stability studies of glyosaminoglycan chitosan complexes", *Electrophoresis, 18* **1997** 745-750.

146. S. Hyase, Y. Oda, S. Honda, and K. Kakehi, "High-performance capillary electrophoresis of hyaluronic acid: determination of its amounts and molecular mass", *J. Chromatogr. A, 768* **1997** 295-305.

147. J. Grimshaw, A. Kane, J. Trocha-Grimshaw, A. Douglas, U. Chakravathy, and D. Archer, "Quantitative analysis of hyaluronan in vitreous humor using capillary electrophoresis", *Elec-trophoresis, 15* **1994** 936-940.

148. J. Grimshaw, J. Trocha-Grimshaw, W. Fisher, A. Rice, S. Smith, P. Spedding, J. Duffy, and R. Mollan, "Quantitative analysis of hyaluronan in human synovial fluid using capillary electro-phoresis", *Electrophoresis, 17* **1996** 396-400.

149. S. Pattanaargson, and J. Roboz, "Determination of hyaluronidase activity in venoms using capillary electrophoresis", *Toxicon, 34* **1996** 1107-1117.

150. N.K. Karamanos, and A. Hjerpe, "High-performance capillary electrophoretic analysis of hyalu-ronan in effusions from human malignant mesothelioma", *J. Chromatogr. B, 697* **1997** 277-281.

151. S. Honda, T. Ueno, and K. Kakehi, "High-performance capillary electrophoresis of unsaturated oligosaccharides derived from glycosylaminoglycans by digestion with chondroitinase ABC as 1-phenyl-3-methyl-5-pyrazolone derivatives", *J. Chromatogr., 608* **1992** 289-295.

152. Z. El Rassi, J. Postlewait, Y. Mechref, and G.K. Ostrander, "Capillary electrophoresis of car-boxylated carbohydrates", *Anal. Biochem., 244* **1997** 283-290.

153. T. Toida, and R.J. Linhardt, "Detection of glycosaminoglycans as a copper (II) complex in capillary electrophoresis", *Electrophoresis, 17* **1996** 341-346.

154. Y. Liu, and J.K.-F. Chan, "High-performance capillary electrophoresis of gangliosides", *Elec-trophoresis, 12* **1991** 402-408.

155. Y.S. Yoo, Y.S. Kim, G.-J. Jhon, and J. Park, "Separation of gangliosides using cyclodextrins in capillary zone electrophoresis", *J. Chromatogr. A, 652* **1993** 431-439.

156. Y. Mechref, G.K. Ostrander, and Z. El Rassi, "Capillary electrophoresis of carboxylated carbo-hydrates I. Selective precolumn derivatization of gangliosides with UV absorbing and fluores-cent tags", *J. Chromatogr. A, 695* **1995** 83-95.

157. Y. Mechref, G.K. Ostrander, and Z. El Rassi, "Capillary electrophoresis of carboxylated carbo-hydrates. Part 2. Selective precolumn derivatization of sialooligosaccharides derived from gangliosides with 7-aminonaphthalene-1,3-disulfonic acid fluorescing tag", *Electrophoresis, 16* **1995** 1499-1504.

158. Ph. Morin, F. Villard, and M. Dreux, "Borate complexation of flavonoid-O-glycosides in capillary electrophoresis I. Separation of flavonoid-7-O-glycosides differing in their flavonoid aglycone", *J. Chromatogr., 628* **1993** 153-160.

159. Ph. Morin, F. Villard, M. Dreux, and P. André, "Borate complexation of flavonoid-O-glycosides in capillary electrophoresis II. Separation of flavonoid-3-O-glycosides differing in their sugar moiety", *J. Chromatogr., 628* **1993** 161-169.

160. U. Seitz, P.J. Oefner, S. Nathakarnkitkool, M. Popp, and G.K. Bonn, "Capillary electrophoretic analysis of flavonoids", *Electrophoresis, 13* **1992** 35-38.

161. T.K. McGhie, "Analysis of sugarcane flavonoids by capillary zone electrophoresis", *J. Chromatogr., 634* **1993** 107-112.

162. F. de Simon, I. Estrella, and T. Hernandez, "Flavonoid separation by capillary electrophoresis . Effect of temperature and pH.", *Chromatographia, 41* **1995** 389-392.

163. C. Bjergegaard, S. Michaelsen, K. Mortensen, and H. Sorensen, "Determination of flavonoids by micellar electrokinetic capillary chromatography", *J. Chromatogr. A, 652* **1993** 477-485.

164. H.-J. Gaus, A. Treumann, W. Kreis, and E. Bayer, "Separation of cardiac glycosides by micellar electrokinetic capillary electrophoresis", *J. Chromatogr., 635* **1993** 319-327.

165. S. Michaelsen, P. Moller, and H. Sorensen, "Factors influencing the separation and quantitation of intact glucosinolates and desulphoglucosinolates by micellar electrokinetic capillary chromatography", *J. Chromatogr., 608* **1992** 363-374.

166. J. Liu, O. Shirota, and M. Novotny, "Capillary electrophoresis of amino sugars with laser-induced fluorescence detection", *Anal. Chem., 63* **1991** 413-417.

167. J.T. Smith, and Z. El Rassi, "Capillary zone electrophoresis of biological substances with surface-modified fused silica capillaries with switchable electroosmotic flow", *J. High Res. Chromatogr., 15* **1992** 573-578.

168. V. Steiner, R. Knecht, K.O. Börnsen, E. Gassmann, S.R. Stone, F. Raschdorf, J.-M. Schlaeppt, and R. Maschler, "Primary structure and function of novel O-glycosylated hirudin from the Leech *Hirudinaria manillensis*", *Biochem., 31* **1992** 2294-2298.

Subject Index

CHROMATOGRAPHIA CE-Series, Vol. 1;

ed. by Kevin D. Altria

Analysis of Nucleic Acids by Capillary Electrophoresis

by Christoph Heller (Ed.)
1997. x, 313 pp. Hardc. DM 168,00
ISBN 3-528-06871-X

Contents: Part 1: Basic Concepts: Separation Matrix - Electrophoresis theories - Microscopic studies.
Part 2: Factors affecting the separation: Electric field and polymer concentration - Sample matrix and injection – Agglomeration.
Part 3: Development in instrumentation: Pulsed field CE - New type of separation matrix - Blotting - DNA separation.
Part 4: Applications: Separation of restriction fragments - Analysis of oligonucleotides - DNA sequencing - Antisense DNA - Mutational analysis

This book on capillary electrophoresis is unique in its focus on the separation of nucleic acids. The importance of electrophoretic separation in every molecular biology laboratory justifies this specialization, which is also reflected in the development of instrumentation. This book is aimed to help to implement this rather new and promising technology in the biological laboratories and to help to overcome typical problems that can occur when starting with a new technique. It should also trigger further development in the field. It covers the theoretical background as well as practical examples of usual applications.
The authors are all experts in their field, having many years of experience with capillary electrophoresis and their application to nucleic acids.

vieweg

Abraham-Lincoln-Straße 46
D-65173 Wiesbaden
Fax (0611) 78 78-420

Stand 1.11.98
Änderungen vorbehalten.
Erhältlich im Buchhandel oder beim Verlag.

CHROMATOGRAPHIA CE-Series, Vol. 2;

ed. by Kevin D. Altria

Analysis of Pharmaceuticals by Capillary Electrophoresis
by Kevin D. Altria (Ed.)
1998. xii, 285 pp. Hardc.
DM 168,00
ISBN 3-528-06696-2

Contents: General introduction to CE principles, instrumentation and methodology - Determination of drug related impurities - Main component assay - Chiral separations - Drug counter-ion determinations - Identity confirmation testing - Clinical applications - Trace level determination of drugs and surfactants - Analysis of excipients and raw materials by CE - Analysis of biopharmaceuticals - Method performance optimization guidelines - Method validation - Critical comparison of CE with HPLC for pharmaceutical analysis - Miscellaneous: electrochromatography, vitamin analysis, radiopharmaceutical analysis, micro-preprative CE

This book describes the current state of the art for the analysis of pharmaceuticals by capillary electrophoresis and contains several hundred references to specific applications and methods. The main purpose of the book is to present the application possibilities of CE an therefore tabulated application data are provided.

Chapters of the book are devoted to providing details of individual application areas such as chiral analysis, determination of drug related impurities, determination of drug counter-ions, drug residue monitoring and main component assay. An introductory chapter provides theoretical background to CE an related techniques. A chapter is dedicated to capillary electrochromatography which highlights the importance this technique currently possesses. Successful regulatory acceptance of CE methods is also described. A comprehensive chapter covers method validation aspects. Other chapters include discrete areas such as the use of non-aqueous solvents, forensic applications of CE, the application of experimental designs, determination of drugs in biofluids, and the analysis of vitamins by CE.

vieweg

Abraham-Lincoln-Straße 46
D-65173 Wiesbaden
Fax (0611) 78 78-420

Stand 1.11.98
Änderungen vorbehalten.
Erhältlich im Buchhandel oder beim Verlag.

A practical introduction to capillary electrophoresis separation techniques

Capillary Electrophoresis

Methods and Potentials
by Heinz Engelhardt, Wolfgang Beck, and Thomas Schmitt
1995. x, 215 pp. Hardc. DM 79,50
ISBN 3-528-06668-7

Capillary electrophoresis combines the analytical separation technique of classical electrophoresis with the instrumental potential of modern chromatographic detection and automation. It eminently supplements chromatography for the separation of polar and water-soluble substances. The extent of its applications is extremely broad, spanning the range of separation from small cations to the highest molecular weight ionic biopolymers. The breadth of its application potential has now made capillary electrophoresis the fastest growing area of instrumental analysis. This book provides a practical introduction to capillary electrophoresis separation techniques. Particular value is placed on the developing and optimizing a separation. Insofar as possible, instructive examples were selected for the explanation of these processes, without recourse to mathematical details. It is hoped that this short introduction will facilitate the beginner's entry and use of this efficient methodology.

vieweg

Abraham-Lincoln-Straße 46
D-65173 Wiesbaden
Fax (0611) 78 78-420

Stand 1.11.98
Änderungen vorbehalten.
Erhältlich im Buchhandel oder beim Verlag.